Opera in the Twentieth Century

OPERA
IN THE TWENTIETH CENTURY
Sacred, Profane, Godot

ETHAN MORDDEN

New York   OXFORD UNIVERSITY PRESS   1978

Copyright © 1978 by Ethan Mordden

Library of Congress Cataloging in Publication Data
Mordden, Ethan
Opera in the twentieth century
Includes index.
1. Opera—History and criticism.   2. Music—
History and criticism—20th century.   I. Title.
ML1705.M67   782.1'09   77-23745
ISBN 0-19-502288-2

Printed in the United States of America

To Dorothy—
ah, let's listen

Preface

This is neither a detailed musicological study nor a comprehensive catalogue, but rather a historical analysis of Western music drama from roughly 1900 to the present. Cutting in so sharply at what is after all a nice round year but an altogether arbitrary date insofar as artistic flux is concerned, and ending with spectacular approximation in the middle, really, of nowhere, this book may at first appear to constitute but a part of a complete set—the fourth, say, of five tomes, or, better, halfway on in a series of uncountable volumes commissioned by a charitable foundation, launched decades ago by our seniors and to be finished decades hence by our heirs. But I trust that a reading of the work will prove its structural integrity and explain why it begins where it does and why it appears to end like the quixotic inhabitant of an animated cartoon, walking happily and unharmed off the edge of a cliff into the ozone.

It is a guide of sorts for the operagoer, to give him some aesthetic perspective on the works that he may already know and to acquaint him similarly with those he does not. The route is, again, arbitrary; still, it has been planned to take in not only the prominent stopping-places from Puccini through Strauss to Britten, but also those less-well-traveled points of interest that are increasingly being cited today in the form of concert performances, gramophone recordings, and festival revivals. It is panorama, images and events.

All translations of non-English quotations are the author's except where noted, and but for the Dancing Master's paragraph out of *Ariadne auf Naxos* and the snippet of *Le Testament de la Tante Caroline*—where it was thought that style and a certain swank took precedence over punctilio—the translations tend to the exact, in the interests of science if at the expense of poetry. Likewise, the spelling of Russian names and titles, which will look odd to most Westerners, is a phonography of the originals rather than that imprecise transliteration of the Cyrillic alphabet usually encountered in English.

In recognition of those who have assisted in this project, I salute the music houses of what is now a past age, when it was business as well as their pleasure to publish the scores of virtually every composition that could decently be called opera, thereby leaving a far-flung library for those who can negotiate the visit; perhaps, too, a word should be put in for the scalawags who man the out-of-print shops of emigré Manhattan, decaying in their starry lofts and mushroomy crannies, librarians, the last of their line. In particular, I wish to acknowledge the strategic advices and queries of Sheldon Meyer and Ellen Royer; I am lucky enough to be surrounded by severest critics.

New York   E. C. M.
September 1977

Contents

I
Induction, 1

II
Allemonde, Shemakha, and On to Peking, 17

III
The Last Days of Romance, 49

1. FRANCE, 51   2. ITALY, 70   3. GERMANY, 106

IV
Jazz, Fugue, Tone Row, and Other Wonderful Motions, 135

V
The Idea of a National Opera, 163

VI
The Heroes of Good Adventure;
or, The Masks Are Down, 209

1. FRANCE, 211   2. ITALY, 222   3. GERMANY, 234
4. ENGLAND, 273   5. THE UNITED STATES, 298

VII
Three Scarecrows and a Pair of Feet, 323

Index, 347

I
Induction

There is always romance, and there is always satire, and as either Napoleon or Talleyrand once said—for no one is certain which of the two it was—"It is only a step from the sublime to the ridiculous." Granted, some periods specialize in the one, some in the other; a few geniuses of the past transcended era and mastered both. Generally, however, there is this tug of war, and from time to time either sublime romance or ridiculous satire will seize the imagination at large and mint its image upon all the issue of the realm, invading the theatres, the waiting rooms of publishers' offices, the exhibition catalogues, and the low media. Now one, now the other: this is the fortune of art.

Opera, in its youth, tended to the sublime. Founded in homage to the old Greek theatre by the poets and musicians of Count Giovanni de'Bardi's Camerata in Florence at the close of the sixteenth century, opera intended to revive the spirit of tragedy in an era when costume pageants and low comedy ruled the Italian stage. Now, it seems that the lines of the Greek plays were sung, not spoken, so if the Camerata were to raise up a Greek drama, it must be a musical drama, for opera is not, as some have claimed, a corrupted theatre—theatre is impoverished opera. Such few remnants of Attic song as the Camerata had to examine convinced them that a monodic chant midway between speech and song would reveal anew the ingenuity of Greek truth; Plato, after all, saw music as first the word, second the rhythm, and only at last the sound, and the Camerata was all for Plato. They thought themselves classicists, and seeing classicism as the imitation of nature, they swore to uphold the sanctity of poetry by rejecting the "unnatural" polyphony that was proving so useful to the writers of the Italian madrigal. It was solo voices for the Camerata, for only the solo—so they sensed—spoke true. The Camerata's was a late mannerist period of art, but they overthrew fashion in proposing simplicity and a certain modified finesse. Legend furnished them their copy, reason their aesthetic, grace their dynamic. They moved, sublimely, in a perfect dream.

At the same time, elsewhere in Europe, others moved in reality, with the aesthetic of the screens and ladders, of the coquette, the student, and the deceived husband, of the braggarts, bumblers, and beguilers: of comedy. Here, of course, was the ridiculous, based not on a Greek musico-geometric "nature" but on naturalism and the indignity of life. Whether in the nice varieties of the *commedia dell'arte*, the *all'improvviso*, the *a maschera*, the *a soggetto*, or in the ballad comedies of the French fairgrounds, the ridiculous concentrated on vernacular art, on the directness and informality of self-conscious popular expression.

The Camerata's theatre was incantatory, devoted to mythic ceremonies of music, to the sharing of humanism, while the ballad comedies traced the pragmatic arc of satire—and these are the polar agencies of musico-dramatic expression. The one is a healing ritual, cathartic, interior; the other is a catcall, upfront and strutting. The gentlefolk invited by the Camerata to hear their private performances were gathered to become one with the vision and to take belief upon themselves. At the fairgrounds, a more democratic assembly never suspended its awareness of time and place; indeed, everything about the ballad comedies conspired to remind them of who and where they were, whether in the dapper asides of the chief comic addressing the audience as if he knew them (why not? he did) or in the familiarity of the folk tunes borrowed for some transparent purpose—for melodrama, sentiment, or tinpot morality. Opera was remote but reachable by a journey of the senses, a giving over and going into; the ballad comedies were already within reach; one didn't even have to turn around.

The historian, casting his eye on all this and ready with 1900 and points thereafter, has one perquisite: he can devise terms. That's what he does. We want terms for these two opposed functions of musico-dramatic technology, because the prime strategy of opera as the years go on is an increasing dependence on the mythic confirmations of music until the verbal absolute is overturned and must actually negotiate a comeback. Interestingly, *logos*, the Word, was the major component of both sublime opera and ridiculous musical comedy—for the Word startled the Florentines, being true (and therefore logical) and the Word armed the comedians, being logical (and therefore, yes, true). But while comedy always hews to the satirical realism of *logos*, opera, seeking romance, will eventually tack to the far side of music's powers of mythic persuasion, and, as the story-ceremonial, will develop its covenantal address of the public. So deeply into music will it then go, that the apprehension of the Camerata, which originally inspired the form we call opera, will have vanished, the Word buried in symphonic metamorphoses.

Romance versus satire, tragedy versus comedy, music versus word, sensation versus intellect, ceremony versus naturalism—all these represent the confrontation of the sublime and the ridiculous. But as the former is often enough pseudoheroic and far from sublime, and as the latter may draw quite convincing and not at all ridiculous distinctions about life here on earth, these words will not do. We call for terms, shorthand to encase the concepts, and the terms to use are *sacred* and *profane,* which the historian can expand to guide us through the complex remixing of aesthetics that occurs in twentieth-century opera.

At root, the rationale for this dual aesthetic of the modern era is based on a wish no longer merely to dramatize but to interpret given characters and their action. The sacred is the mode of dramatization—which is why poetry and music, the great dramatizers, are crucial to its aims. The profane, however, is intrinsically interpretive, too aware of its own dynamics to want to cast a spell; it engages its public by direct address, cutting off any possibility of the almost eucharistic exchange of the sacred and its musical offerings. Thus the spoken language, especially in the vernacular, is a gambit of the profane, for nothing is so useful as speech in the musical arena to check the flow of fantasy, to jar the spectators out of any reverie they might succumb to under the narcotic of music. Doubtless the best example of the sacred is Wagnerian music drama (whether Christian, pagan, or, as in *Die Meistersinger,* middle class), with its mesmerization of the spectator through the sheer power of its music. The most representative example of the profane would be a musical comedy with the Marx Brothers, such as *Animal Crackers* (1928), wherein the autotelic theatrical frame is repeatedly broken by parody based upon wild allusions, a facile musical technique, and Groucho's blithe grousing at the audience.

Two such polar examples would seem to refute Napoleon/ Talleyrand's statement about there being only a step from the sublime to the ridiculous—but remember, it is not subject matter but musico-dramatic machinery that defines the aesthetic. It is indeed only a step from the one to the other, sacred to profane

and back again, and by the twentieth century the two will begin
to collaborate in opera. A freespoken *res publica* of form is the
goal, a revolution; we will encounter this form when the neoclassical revival of the 1910s imposes the logic of comedy, through
the Word, upon the prevalence of music, yielding the best of
both romance and satire while chawing the bone of myth. And
this is not, after all, a new idea: Greek tragedy was the fount of
sacramental presentation, but the Greek festivals did not neglect
comedy, and such awesome undertakings as Aeschylus' *Oresteia*
regularly melded a trilogy of crushing gravity with a fourth
piece, the satyr play, on the same storyline but in a prankish
vein. Opera had been comparably ecumenical in its Venetian
heyday, in Monteverdi's era, but by the eighteenth century it
had undergone purification, the satyr's descendants forced off to
the side. All over Europe, on the outskirts, within the shadow
of the city walls come fairtime, the outlaw and unshakeable profane complemented the sacred of the inner-city theatres. While
opera seria developed the popular taste for the grand gesture,
however hollow, French vaudeville, German *Singspiel*, English
ballad opera, and Italian *opera buffa* filled in with the realities
of life and proclaimed the spoof. At any given moment the hip
gentry of Paris had two shows to take in—the latest *tragédie lyrique,* and its burlesque counterpart at the local vaudeville house.

In the 1750s, all Paris had been captured by a piece on upward
mobility, Giovanni Pergolesi's *La Serva Padrona—The Maid
Turns Mistress.* Its riotous success precipitated a ferocious pam-
phlet war, the celebrated "querelle des bouffons," presumably
upholding French music (as originated, however, by the Italian
émigré Jean-Baptiste Lully) against the Italian Pergolesi—but
actually defending pseudoheroism against comedy, and no wonder the former lost out. The sacred, at its worst, is a magniloquent

charade, a tale that fails to tell, while the profane is ever on the beam, much closer to its audience. From the success of *La Serva Padrona* emerged French comic opera, which turned out to be the ballad comedies of old but without extemporization and anonymous tunes, while still retaining their directness. Outward from Paris radiated the news.

With vernacular comic opera set on its feet, the symbiotic interplay of sacred and profane, music and text, myth and Main Street—a dynamic of opera throughout its history that becomes an overriding aesthetic in the twentieth century—was ready to begin. It remained only for some messiah to redeem the comatose pageantry of *tragédie lyrique* and *opéra-ballet*, the dancer's ambit, and the more agile but equally excessive *opera seria*, the singer's arena, with the verity of drama. Just such a man was Christoph Willibald von Gluck, who, in the company of some talented librettists, pointed the way back to the splendid dramatic contraption of the Greek theatre and at the same time liberated the dramatic orchestra. With Gluck—and this is crucial to the fortunes of the sacred—the music in music drama was neither the servant of the words nor its separate but equal teammate, but the very core of the action, the leader, the teacher. Violating the fashions of the day, Gluck's music soon had its opposition, which in turn found a champion in Niccolò Piccinni, a worthy of the Neapolitan school but no champion. Neither Gluck nor Piccinni acknowledged the rivalry that Paris arranged for them, yet the affair came to a head when each ended up composing an *Iphigé-nie en Tauride* to different librettos. Gluck's version, in 1779, an outrageously passionate adaptation of Euripides, was a precursor of the sky-hurtling impulsions of nineteenth-century romanticism, while poor Piccinni's *Iphigénie* two years later sounded shallow by comparison, and this round went to the sacred.

Why all this now, this .. . . . this history? Bring on Louise and the Unknown Prince, the Marschallin, Doktor Faust, and the Bassarids! Let there be turns of the screw and mothers of us all! What have Gluck and Piccinni to do with it? Everything. In order to comprehend what the modern Faust is up to, one must first crack the code of form; form is the modern romance, the new intelligibility. And the years between *Louise* and *The Bassarids* are all for form, a form musico-dramatic in structure, as ever, but increasingly word-charged and parodistic in insight, haunted by logos, the sacred and profane in confrontation. That is precisely what that duel of *Iphigenia*s amounted to—thus the review—for like the *querelle des bouffons,* it pitted Gluck's chivalric music drama against Piccinni's vernacular comedy. The former won, and Gluck's symphonic actuation of opera was to rule for the next century.

Romanticism recognized chaos—practically invited it—and demanded true exploits. But the romantics had little sense of satire, of the nihilism of comic rhythm, and the last great moments of the profane before the twentieth century belong to Mozart and Da Ponte. Note that their *Don Giovanni* was announced as a *dramma giocoso,* a joyous drama that plays heroic archetypes off comic archetypes: only moments after the curtain has gone up, when Donna Anna and Don Giovanni are wrestling in fearsome earnest, Leporello games it up under their legato thirds with a bass line of pagan patter.

The absurd seriocomic world of modern music drama tips its hat, ever so slightly, to Giovanni Busenello's sly rape of system, back in 1642, in his libretto to *L'Incoronazione di Poppea,* the first axe blow for the profane in the sacred wood, but it owes a real debt to Da Ponte, and as for Mozart, it was he who proved that the same genius propounds the sacred and profane both. As the years passed, there were few if any composers who didn't attempt a comedy. Wagner and Verdi both brought the form to heel after early comic failures, but the consistency of style of *Die Meistersinger* and *Falstaff* is in one sense less remarkable than the intersection of modes in Mozart's *Die Zauberflöte*—Masonic fraternity and earthy amorality at the junction point. For notice that Mozart was able to stabilize the introduction of Tamino and Pamina's sacred world to the profane one of Papageno entirely through music, world within world. The soprano's "Ach, ich fühl's" gains in its power to move at each hearing, while Papageno's entrance song, sprung from the folk tradition of the fairground ballad comedies, proves comparably fetching on a more informal level; the one leads us on to heights, while the other has been fast with us all along. But when the two worlds collide, Mozart can set Papageno and Pamina a duet, "Bei Männern, welche Liebe fühlen," without forcing a note or sacrificing either character's resonance.

Few were the librettists who utilized this interrelation of poetic interior and comic exterior in the nineteenth century, and between the times of Mozart and Busoni the heroes dwelled in literature, not librettos. Except for Berlioz' *Les Troyens,* an epic on heroic destiny that was rejected by its own time, Wagner's redemptive legends, and Verdi's melodrama—the most virile theatre music since Elizabethan verse—the nineteenth century specialized in Gothic horror, languid love-and-duty, and historical potpourri. But it was popular art, one that attracted a broad following. Music was dominant. Every opera was full of Big Tunes and a melodic line that stirred the emotions. Not truly heroic, though it thought it was, nineteenth-century opera more often attitudinized than achieved, and made too few real artistic demands. One hears of the *aria di sorbetto,* sung by some dingy comprimario while the house gorged itself on sherbet and scandal, but the Italian bel canto era of Bellini and Donizetti's time brought forth whole operas *di sorbetto*, at least to modern eyes and ears, so bird-brained are the words and so placidly arabesque the music. Best-selling novels and hit plays were a favorite source of opera plots, so it is fitting that when librettos once more became important at the end of the century, plays would simply be set to music with a slight cutting of lines.

In the nineteenth century, however, most composers felt impelled to reach the public not by drawing it into a work, but by extending a work to meet the people. This was a nationalistic era. As far back as Haydn, the folk had been making itself heard in music, especially in symphonic dance movements, and few were the rescue operas of the late eighteenth century that did not find room for a secondary, folk-wise couple, as if in response to revolutionary developments in America and France. Similarly, after Gluck, opera had become something of a phenomenon; ingratiating airs were whistled in the street. But while music increasingly dominated the drama in opera, the result, with rare exceptions, was lackluster nobility that fell far short of the concept of the sacred which Camerata had attempted to re-create from Greek drama. Even lean and eliciting Verdi, more responsive to the Word, as both dramatic mechanism and reflection of national aspiration, than any other composer of his time, flubbed his early attempt at comedy and did not successfully deal with the profane elements until his final triumph in *Falstaff*.

Of course, the nineteenth century was a heyday for music drama—such grand operas!—whatever its slant. Verdi and Piave did deal with the contemporary Parisian demimonde in *La Traviata*—yet for all its intended naturalism, *La Traviata* did not mix the profane with its intense drama: music did the work. *Opéra comique,* however, with its spoken dialogue, closed forms, trivial musicality, and intermittent moral, sentimental and comic splashes, could treat a premise similar to that of *La Traviata* in a vastly different way, as Daniel-François-Esprit Auber did in his *Manon Lescaut,* a close coeval of *La Traviata. Manon Lescaut* ends with a famous death scene unexpectedly grand for the form, perhaps approaching the pathos of *Traviata's* final act, but on the way there its structure, postures, and personality adhere more to the tenets of comedy than "heroic" tragedy. In Auber, as not in Verdi, minute but unsubtle jolts bring one down to earth each time the music comes to a stop, and this is meant to be. One is not enthralled; one does not believe; one enjoys, but one remembers always the artifice in the scheme. Begging for a pass to visit her true love at his barracks, Manon can with aplomb address her temporary protector in a measured phrase complete with notable trill, be answered by the protector to the same music, trill and all, then hear his seductive waltz in response, and at last join with him in a lively finale and stretto, all of this indicating but not transmitting great urgency.

How differently Verdi handled such scenes—how different, in other words, the sublime from the ridiculous—with less "realism" but greater art, for musicality is endemic in everything he wrote. In Verdi the music pulls one into the scene, forcing one to align with the worldview for a moment; in Auber, conversely, one is alert and makes no voyage. This is not to say that Auber's method has no power as such, but that its impact is more intellectual than spiritual. The art of opera, with its antithetical concatenation of action and expression, may stimulate its audience either as quick-witted thrusting out (the profane) or as sensual drawing in (the sacred). The historian does not invent; he only describes.

With Wagner, we come to the apogee of the sacred, for Wagner's music drama is built out of a musical will, however thrilling the librettos, and relies on myth to ignite the imagination even as it suspends the mind's disbelief. Wagner brings us closer to Debussy and the twentieth century, too, for Wagner's sacred exhibit, toned down for projection of the Word, is the format of *Pelléas et Mélisande*. As early as 1850, Wagner decreed the explication of myth as the poet's duty, and he validated his stand by aligning the inherited apprehensions of the Western unconscious with modern philosophy, amortizing old myth with new myth. On the musical side, Wagner collaborated with Schumann, Berlioz, and Liszt in devising a cyclic expansion for the music in the drama, raising it to a pitch of contiguity that made the libretto something to be studied at leisure rather than grasped on the spot in the theatre.

With the *Ring* and *Tristan und Isolde*, Wagner the poet was eminent enough to convert the operagoer to his vision, and in this he incontestably rerouted the direction of operatic storytelling. Thomas Mann, among others, hailed Wagner as the "savior of opera" for his resurrection of theatre myth, the story-ceremonial, noting with a polymath's inference the broad folk-cultural perspective of a scene such as Siegfried's death march in *Götterdämmerung*, "the first and farthest of our human picture-dreamings"—Tammuz, Osiris, Adonis, Prometheus, Diarmuid, Christ . . . the kings who must die, "the whole world of slain and martyred loveliness this mystic gaze encompasses."

Heavy stuff? Some thought so; whistles, diatribes, duels, and Nibelungen spoofs in the 1870s rivaled the Gluck-Piccinni battle, and for precisely the same reason—all this vaulting heroism tested the public patience. For recovery, for refreshment, for escape, it unwittingly patronized not the antidote but the *com-*

plement to Wagnerian sacrament, satiric comedy. No composer worth his salt has failed to be awed by what Mozart was able to do with the profane in his operatic masterpieces, and by the late nineteenth century, vernacular comedy had turned to satire, and here the realism of the Word counted for all.

This is where the profane largely resided in the 1870s and 1880s, in stories looted from *opera buffa,* but with strong popular elements in its humor and its musical roots. The "genre primitif et gai" of old *opéra comique* that Jacques Offenbach announced as his objective has but one tenet—that of spoken dialogue alternating with musical numbers, rather than through-composition— but it carries a host of presuppositions, such as simplicity, clarity, balance, pace, and irreverence, which in the aggregate produce a form all its own, useless for sacred idealism but excellent for satire. *Le gai* comprises any workable felicity, but *le primitif* is a special component, invoking what is most basic and integral in life; this element, doubtless, has attracted twentieth-century opera, for opera was never so contemporary before Ernst Křenek's Jonny cut a pose and struck up his impervious *Jazzband* in the 1920s.

This *primitif* is a curious beast, simple, simple . . . and yet it is the foundation of modern innovations in music drama, complex alike in the harmony of their precepts and the point of their melody. Ferruccio Busoni will build his, the "other," *Turandot,* before Puccini's, on the memory of authentic Venetian commedia, but in sounds of such adroitness that the folkish C Major of the finales falls on ears made young enough to hear freshness in the diatonic. On one level, the subtlety of twentieth-century opera lies in its unleashing of the Word from its bondage to music, a recognition of the arrival of satire; but it is how the music supports the libretto that carries the impact. We now enter the twentieth century, the era of seriocomedy, when myth merges with *le gai primitif*, a time of Wozzecks and Don Quixotes, women without shadows and girls of the west, dinners of jests and elegies for young lovers, prisoners, dybbuks, gamblers, and noses, of obsessed Baby Doe and wan, bewildered Vasco the barber.

It is 1900. Wagner's authority is secured, and the reconnection back to Greek tragedy is, in a loose sense, accomplished—but music, not text, is king. Verdian melodrama has evolved into the fleet music drama of *Falstaff*, the format of which might work forever but for the mischance that no second Verdi is going to surface to work it. Naturalism and romance have overthrown the Word for the gran duo with its ineluctable Big Tune. Arioso and cyclic unity are preferred to spoken dialogue and recitative. Music has passed into dramatic law.

Well, we'll just see about that.

II
Allemonde, Shemakha, and On to Peking

After a century of operas named after such heroines as Agnes of
Hohenstaufen, Jaguarita the Indian Maid, the Lady of the Lake,
the Queen of Sheba, the Snow Maiden, the Daughter of the
Regiment, and the Vestal, *Louise* inaugurated the twentieth
century like a ha'penny pastry subverting a bakery—plain Louise,
a dressmaker. The public had long since submitted to the proposition that not every diva could be a princess, or even a princess
disguised as a slave, but never had the world of urban sprawl
been presented before it so completely as in this "musical novel,"
as Gustave Charpentier termed his piece. Acting, with a little
assistance from others, as his own librettist—or, rather, as his
own composer, given this musical novel—he caught in *Louise* the
daily life of daily people as no one had yet attempted to do, or
thought useful to attempt.

Naturalism begat *Louise*. Fleeing from the regal excess of romantic opera as well as from the vapid brio of domestic comedy,
the naturalists read up on their Hugo and their Balzac and rallied
'round Emile Zola, who preached the drama of environmental
predestination. Whether in the city or out in the countryside, the
folk were bound by heredity, custom, and habit to do what they
did, and recognition of this phenomenon gave rise to a school of
opera composition in Italy called "verismo." This has been
widely misapplied today to a whole generation of Italian musicians committed to fantasy and visionary symbolism, whereas

verismo actually endured a brief run in the 1890s and early 1900s, heralded by *Cavalleria Rusticana* and *Pagliacci* (both of whose composers were almost immediately graduated to legend, love-death, costume epic, and other wonderful motions). Devoted to explosive earthiness, verismo favored the passion of the mime without its poetry, and by its very nature could sustain neither a long evening nor a long era.

Louise, however, found the luster in quotidian calm. Its characters get angry, as people will, but they don't enter gnashing their teeth and exit committing murder; they take their time, letting the action progress to a conclusion neither happy nor doleful, but, as in life, open-ended, to be continued. Unlike verismo melodrama, *Louise* contains few climactic moments or confrontations, and it's a long piece, almost never performed in toto. Shorn of instant thrills, it grows on one slowly, but its simple charm is instantaneous, and its first night, February 2, 1900, was a triumph.

There were a few holdouts. Ambroise Thomas, the Napoleon of the Conservatoire, had written Charpentier off when Thomas detected the quotation of a popular tune in Charpentier's tone poem, *La Vie du Poète*—popular is not art—and there was much more to affront him here, particularly the scene in which Louise and her parents sit down to a frugal working-class dinner. What, *musique de potage?* Others, less musical, thought *Louise* a glorification of vice, but the younger generation accepted it as their carmagnole—Paul Morand dubbed it "the socialist's *L'Aiglon.*"

Perhaps the first opera to be set in Paris—truly set in it, fretful with the tense, tempting aggression that has failed to enchant no one since the Vikings attempted to adopt it in 885—*Louise* features an enormous cast of street people and bohemians but closes in on four principals—Louise, her father and mother, and her lover, the poet Julien. Charpentier weighted his text heavily against the parents, who oppose their daughter's romance and end only in driving her away. But how could it stand otherwise in a work by a man who extolled *The Life of the Poet*? The demimonde haunts the opera, calling to Louise even through the walls of her parents' tenement flat and amidst her chattering fellow dressmakers at work. "Free!" it shouts, egging her on in her duet with Julien in Act III, just after her solo, "Depuis le jour," a bizarre piece that manages to sound simultaneously chaste and feverish. It is Julien who speaks for Charpentier. "Everyone has the right to be free," he says, "and every heart the duty to love! Bad luck to him who would strangle the proud, individual will of the soul that awakens to take its share of the sunlight, its share of love!"

No wonder *Louise* when it appeared looked like the fountainhead of a new epoch. Its immediate predecessors in French opera ran to the likes of Jules Massenet's *Cendrillon*, Ernest Reyer's *Salammbô*, and Vincent D'Indy's *Fervaal*—the elegant, the sultry, the somewhat Wagnerian. *Louise* made real opera of the realism that French novelists had discovered for fiction in the 1830s—the singers were dressed in clothes they might easily have worn on the street—and Charpentier's music, too, seemed fresh and impetuous, as if caught on a binge: the multihued preludes to Acts II and III, "Paris Awakes" and "Toward the Distant City," each a mosaic of stirrings and lucky strikes; or the intermittent street-vendor's cry, "Régalez-vous, mesdam's," transformed into the jubilant anthem of the bohemians at their party in Act III; and that waltz, that quintessentially Parisian expression of cheap loveliness in three-quarter time. Charpentier rams the tune home in the last act, when the enticement of the city drives Louise half-mad, but her friend Irma sings it first, in the dressmaking workroom, over a hubbub of chatter, for in this opera, as in life, someone might well interrupt one's nicest invention. The gossips, however, soon take notice of Irma's eloquence and subside just in time for the waltz tune. "A mysterious voice," says Irma, "in the rustling of the friendly street, promising happiness, pursues me and coaxes . . . it's the voice of Paris!"

Louise's parents take a dim view of all this. Her mother defines *la vie de bohème* as "misery set to songs," and the confrontation leads to a brilliant fourth act. Louise has come home because her absence was killing her father, but now she chafes at the confinement, impatient with his manipulative authority, and to soothe her, he rocks her on his knee and sings a lullaby. Uneasy, Louise tries to get away from him, but he holds her fast, cradling her as if she were an infant while the orchestra plays a touching little theme harmonized with disturbing altered notes. This vignette sums up the case against the bourgeois enemies of the poetic life, and even when Charpentier lures the slumber song from g minor to G Major and sweetens the string tone, one is revolted—not by the machinations of evil counts or Druid priestesses, but by the sad importunities of a workingman's life. No sudden death, no curse fulfilled blights the denouement: angry at her pouting, the father throws Louise out of the house, and as she flees, the lights of the city go dark. Now he comes to his senses and calls to her, but she is gone, and the curtain falls where the novel would have ended.

Louise, which set the Left Bank at a right angle to the laborer's suburb, gave naturalism its most effortless musical expression —cornered the market, in fact, and so closed the account books, for in opera naturalism could go no further than this, and the next epoch had already made other plans. This is the end, this

Louise, not the transition. There is a line in Igor Stravinsky's *L'Histoire du Soldat,* depending on the translation one takes, about "not sharing what you are with what you were," an epiphany perhaps best realized by a Russian émigré sitting in Switzerland. Music requisitions its continuity, yes, but not in the case of Charpentier's bohemian lovers. A signet of the times, part leftover love music and part new-style conversation piece, *Louise* is 1899 more than 1900. 1900, the real such, is . . . well, let it speak for itself:

"I can no longer see the sky through your hair," he says, two years after *Louise.* "You see, you see my hands cannot hold them anymore. They almost reach the branches of the willow tree. They live like birds in my hands, and they love me, they love me a thousand times better than you do!"

It is Pelléas who speaks, and it is with him that twentieth-century opera truly began, in language musical and verbal that overwhelmed *Louise.* No one saw this more keenly than Charpentier, and he was prepared to accommodate the new mode, turning, in his only other opera, to Louise's lover, but he could please no one with *Julien* (1913). For the laborer's daughter, naturalism; for the poet, mystery. From the rues and ateliers of his first opera, Charpentier invaded the cul-de-sacs of metaphysical quest in his second, a dream play of the sort that led the trends of the day, careening from hallucination to recognition, from Christian parable to the side-show intoxication of Paris at carnival time. Louise was on hand again, briefly as herself and then reincarnated as visions of beauty, innocence, age, and lust, but *Julien* was subtitled "the life of the poet," and so it was. This was Julien's outing, diametrically opposed in form to the literal Louise, and nothing like a sequel, though it is often assumed to be one.

Julien won no favor, and died, but it was an offspring of the era no less than *Louise* had been thirteen years earlier. In the meantime preceptive fantasy was born, and now the examination of critically pivotal operas can commence. *Louise* was simply the last opera of the last century; the first of the next, which changed everything, came to light in 1902 in the same theatre—in Paris, of course, in the twilight of its Holy Roman influence. The new piece was Claude Debussy's *Pelléas et Mélisande*, a setting (with a few scenes omitted) of Maurice Maeterlinck's play of 1892. No adaptation was made. Debussy simply set the text to music.

In clear, simple diction, Maeterlinck's drama resuscitated the Francesco-Paolo situation from Dante's *Inferno*, placing it in a gloomy, feathery, storybook kingdom he called Allemonde. Lost in a forest while hunting, Prince Golaud finds Mélisande, a pallid creature generally incapable of direct statement. Married to Golaud, the girl is attracted to his half-brother; when Golaud catches them embracing, he kills Pelléas and pinks Mélisande, who dies not so much of the wound as from the rigors of childbirth and *mal d'esprit*. The language is ever concise; the action is permanently ambiguous, as if enacted behind gauzes. Think of the palpable candor of *Carmen*, of *Samson et Dalila*, of *Thaïs*, and consider the contract binding author and public; that contract must be renewed every so often, and the terms are always renegotiable. Maeterlinck in particular was a hard bargainer, obsessed as he was with the edgy cut of the puppet play, fashioned of a flesh so unmortal that he feared live actors would wreck the mood.

This idea was reconnoitered constantly throughout the 1900s, perhaps most basically in the theatre pieces of Michel de Ghelderode, sinister farces and mercurial Gothics such as *Sire Halewyn* and *Escurial* (both of which subsequently became operas), and in such a ballet as Stravinsky's *Petrushka,* in which the lifeless agility of the dancer patterns the hollow music of the stringmaster-composer. Late German romanticism, in its fascination with automatons and doppelgängers, left an impression in some quarters that drama called not for the figures of everyday familiarity but rather for changelings, mute or jabbering without reason, forcing the wildest surmise upon the audience by ridiculing the very idea of realism, even of real life.

This is the source of modern seriocomedy, playing against sensitivity by affecting none itself, and opera, with its twofold alembic of poetry and music, will prove redoubtable in the field, even if it can be very bashful about yielding to transformation. Composers and librettists, as opposed to dramatists and novelists, are often slow to adapt from the way we were to the way we are going to be. They were late with symbolism, and then, when Arnold Schoenberg was ready with expressionism, the public was not . . . and even the Florentine Camerata, which more or less founded opera more or less on the classical plan, didn't get around to its share in the classical revival until classicism was fading into baroque.

It didn't matter, though, that Debussy was a bit slow in getting Maeterlinck's world onto the operatic boards, for the score was so ingeniously oriented to the script that *Pelléas et Mélisande* looked to be the penicillin of music drama. In truth, it was no innovation, but an innovative return to an old idea, the Camerata construct, putting the music at the service of the text, emphasizing monody in the vocal line, avoiding anything that didn't belong there initially—namely an aria—and keeping, in its tremulous distance, a sense of discovery and release.

It was Debussy's music, actually, that so stirred audiences to a sense of wonder, for the action of the play is as opaque and undiscovered as anything ever will be. Of all the characters, only Golaud asks questions and wants things spelled out for him; everyone else dwells in uncertainty. At Mélisande's deathbed in Act V, Golaud asks her if she betrayed him with Pelléas. No, she replies, but this is no answer, and Golaud knows it, for Mélisande's no is yes, her yes is perhaps, her perhaps is why, and she has no because. The play, of its indistinct, plangent species, is a kind of masterpiece; Debussy's operatic setting of it is certainly one, a masterpiece whose effects are still being felt today in regard to the mating of words with music.

A former ardent Wagnerian (a little less ardent as the years went by and his own opera revealed itself), Debussy leased old Klingsor's patent for through-composition. Each of the fourteen scenes is a separate musical movement, bound to its fellows through the set changes via musical fantasias, and there is no full stop until the end of an act. There, however, the resemblance to Wagner ended. Debussy had grown fatigued to distraction with operatic formulae, especially those Wagnerian leitmotifs that all but roll over and play dead when the action gets hectic; "Music," said Debussy, "is made for the inexpressible. . . . There's too much *singing* in opera." The eccentric Eric Satie had also considered setting a play of Maeterlinck's, *La Princesse Maleine—Pelléas* as well—and Debussy later credited Satie with having defined the aesthetic for a Maeterlinck opera: "Why should the orchestra grimace whenever a character comes on stage? Look—do the painted trees in the scenery grimace? What's needed is a *musical* scenery, a musical environment in which the characters move and talk."

Doubters wave away this musical scenery as aimless and intermittent, but for the convinced, what wonderful esoteric plunder is to be got here, what long-lined beauty and what—despite many commentators' remarks—power! As if to prove how bashful the piece is, how anti-Wagnerian, the lovers' vows in Act IV (Pelléas' "je t'aime" answered by Mélisande's "je t'aime aussi") have been made exhibit A in the case of the revolt against Wagner, for the exchange is delivered in simple recitative without aid from the orchestra. But this is to ignore the desperate love music that wells up immediately afterward. No, this is no retort to Wagner, but the retort to strutting romantic opera, made in chamber-Wagnerian terms. The composer's marking for the fateful first measures of *Pelléas et Mélisande* is *très modéré*, but what could feel less moderated than this vibratile cave-, water-, forest-, tower-sounding score, with which Debussy seemed to hold all fairyland in fee? Floating his piece on a field of thematic interchange, he chose to begin with a tiny monastic chant alternating d minor with C Major, then gave the woodwinds an almost tuneless melody on two notes in the altered chromes of impressionism: the seed of art nouveau planted in the furrows of legend.

Such was the age that Faust and Don Juan were little heard from—at first—but Pelléas was everywhere. Both Gabriel Fauré and Jean Sibelius wrote incidental music for productions of Maeterlinck's play, respectively just before and just after Debussy's opera, and Arnold Schoenberg's mammoth tone poem on the subject arrived in 1905—and such was the age that, although the escort of distinct forms would do for Fauré and Sibelius, Debussy and Schoenberg delivered Maeterlinck into an organic webbing of sound, a logical completion of the motto themes and inchmeal devices of the romantic symphony. Such was the belief of the musician, that music could stimulate without screaming, and such was his mode that atmosphere, not word coloring, became the goal, throwing the Word into relief and making it, if less passionate, more direct.

Schoenberg was to devise the twelve-tone system of atonal composition more or less singlehandedly, but Debussy plucked at the wiry, horizontal strands of atonality in his score for Gabriele D'Annunzio's play, *Le Martyre de Saint-Sébastien*, in 1911, just two years after Schoenberg's epochal essay in future shock, the Fünf Orchesterstücke. Looking backward, one sees the foundation of much of this in *Pelléas*, and this drift away from tonality—and with it, public approval—will be noted as the years pass, for it signals a new point of view on the composer's part, a giant step to one side of decoration into objectivity, and from objectivity into satire, passing through a mirror, as it were, to glance back and gape, disquieted. If Debussy was to overthrow the grimacing orchestra of romantic opera, fighting fashion with the artillery of art, a scientific extension of his plan resulted in the opera that sounds fixed in a grimace, that travels a route contrary to that of the stage action, plainting as much its own agonies as those of any fictional characters.

The Parisian public cannot be said to have welcomed *Pelléas et Mélisande* by acclamation, and with a few exceptions the music world proper acted as does any establishment agency when confronted with something new and wonderful—in total panic. Critics, mystified as to whether the piece was derived from Wagner or counter to him, but sure it was one or the other, ranted and floundered in their columns, and the Conservatoire solemnly promised to expel any student who heard it. Debussy had further to contend with the anger of Maeterlinck, who had determined that the operatic Mélisande. be created by Georgette Leblanc, the original stage Mélisande and his very good friend. The role had already been assigned to Mary Garden, however, and Maeterlinck demanded satisfaction, first with a spurious challenge to a duel, secondly with an equally spurious lawsuit, and then, more sincerely, in a newspaper campaign dedicated to the opera's swift failure.

This, as the Icelanders used to say, was not to be looked for. *Pelléas* caught on immediately, though in a highly peripheral way, and it remains the least frequently encountered gem in the jewel box of opera standards. Leblanc did sing a Maeterlinck heroine only five years later, in 1907, in Paul Dukas' setting of *Ariane et Barbe-Bleue,* by which time Leblanc had become Ms. Maeterlinck. *Ariane,* another enigmatic fable set in the nether world, got from Dukas pretty much the Debussy treatment: a protean, kaleidoscopic accompaniment running under Maeterlinck's script, an excelling articulation of the text, and the subterranean modality of impressionism, instrumentally describing water, light, darkness, even fear—in short, that "musical environment in which the characters move and talk" that Satie had recommended to Debussy.

Ariane, however, is no puppet play. Subtitled "The Useless Deliverance," it tells how Bluebeard's sixth spouse discovers his first five wives imprisoned in a vault of the castle, offers to release them, and finds that they prefer to stay put in darkness. Interestingly, the wives who elect to remain cut off from the world—Sélysette, Ygraine, Mélisande, Bellangère, and Alladine—are all refugees from earlier Maeterlinck works (Mélisande's appearance invokes a quotation from Debussy's score). They represent Maeterlinck's own feeling that his puppets were leading him into a spiritual miasma, that a way out of his poetic anomie must be found—a way located by Ariane at the close of the piece when she abandons Bluebeard and his wives for the outside world. She is a novelty in Maeterlinck, a votary of action who takes initiatives that would never have occurred to Golaud's wife. "What is permitted," Ariane explains at one point, "teaches us nothing."

Debussy's alternative offered intimate subject matter and a

more accessible communion between performer and public. Nobody could beat Wagner at the mounting of sacramental theatre, or even rival him, but many were trying to, particularly in
France. Impressionism as musical dramatization struck some
critics as the worst sort of mannerist decadence, with its ambiguous harmonics and meandering melody—the Germans, mainly,
scored it as an aberration of fin-de-siècle, though they were to sit
still for and even enjoy the far more outrageous *Zeitoper*—but
impressionist opera proved a Promethean boon to the dramatic
end of things. It was as if Debussy had come down to man with
the fire of textual integrity, yet without losing hold of the unity
provided by germinal themes that imitate and interconnect.

"Men may rise on steppingstones of their dead selves to
higher things," quoth Tennyson, and so it is with art, where
innovation observes tradition and genius is the assimilator who
transforms. The fresh aesthetic that gave us *Pelléas et Mélisande*
and *Ariane et Barbe-Bleue* pays fealty to the Florentines of the
Camerata, to Dargomizhsky and Mussorksky, all making metre
bow to nature by singing the way one might speak if speech
were song. But the Word demands new perquisites from drama,
and where once Jules Barbier and Michel Carré shredded
Goethe for Gounod and Thomas and assassinated Shakespeare
for Thomas' happy *Hamlet*, now Debussy and Dukas simply set
their Maeterlinck originals, slightly cut, to music.

The Maeterlinck era, mannerist or not, was regarded as a
revolt against naturalism, but these artistic epochs have a way
of overlapping, and 1907, for example, the year of *Ariane* and
Frederick Delius' *A Village Romeo and Juliet,* in a kind of folk-
impressionist style, also saw the arrival of Massenet's verismatic
Thérèse, one somewhat sensational hour of love, death, and the
French Revolution. Heading farther east to forbidden Russia,

1907 also hosted the premiere of Nicolai Rimsky-Korsakof's *Ska-zaniye o Nyevidimom Gradye Kityezhe i Dyevye Fyevronii* (*The Tale of the Invisible City Kityezh and the Maiden Fyevronia*), a tremendous sally into the folkish ceremonial and the nature fable, aerial, earthy, awesome, and wonderful.

Isolated Russia, cordoned off from the West long before the Soviets, didn't get a national opera tradition going until the mid-nineteenth century, and as recently as 1890 Chaikofsky was still mining romantic material in *Pikovaya Dama* (*The Queen of Spades*), using closed forms, recitative, and a ghost story derived from Pushkin. It was Rimsky who brought Russian opera into the twentieth century, usually writing the words as well as the music, and varying his musical tone to fulfill his experiments in genre. Unlike Verdi and Wagner, who each refined one essential form in multiple installments, Rimsky zigzagged in mode. For *Mozart i Salieri* (*Mozart and Salieri*, 1897), he exploited classical pastiche; for *Tsarskaya Nyevyesta* (*The Tsar's Bride*, 1899), he brushed the blood-and-thunder canvas of middle-period Verdi, then played diatonicism off chromaticism for the Manichaean man-against-the-supernatural premise of *Kashchey Byesmyertni* (*Kashchey the Deathless*, 1902).

It was the world of legend that brought out the best in Rimsky, as it did in Wagner, and the Russian made his all-out effort in the tournament of mythic opera with *Kityezh*, a four-hour fairytale set in the thirteenth century, during the Tartar invasion of western Russia. Throughout his career, the composer fascinated himself with the symbolic character of pre-Christian man's adaptation to nature and the idealism of pagan belief. His "opera-legend" *Sadko* (1898) delved into the mystery of the eternal sea, giving a liquid princess the top line of an extravagantly harmonized love duet and rounding off the entire score with an undulating motive representing the ceaseless progress of the oceans. Now, in *Kityezh,* Rimsky moved into the forest preserve in search of the sanctum of Mother Russia and her secret City of God. Kityezh's twittery *Waldweben,* saintly heroine Fyevronia, and Orthodox mysticism have nominated it as "the Russian *Parsifal*" in the chronicles, and indeed its association with Wagnerian myth-drama is felt in every scene of Vladimir Byelsky's libretto. But where Wagner imposed a sometimes unnatural language upon his characters, locking them into the score as human instruments, Rimsky celebrated the Slavic inflection in his vocal parts—a procedure native to French music theatre as well as that of the Slavs.

It is in its subject matter that *Kityezh* pointed back to Wagner's mythic construct for opera, and forward to the revival of symbolist legend in the 1910s and 1920s. In his graduation through operatic types Rimsky delivered the mythic valedictory when he arrived at *Kityezh,* his penultimate stage work. Tracing the romance between a Russian prince and a hermit maiden in territory beset by barbarians, Rimsky painted a musical icon, not without its vernacular intrusion in the person of a drunken coward who betrays his countrymen and goes mad in the forest, where nature is in harmony with the patriotic/religious and in opposition to the faithless.

It is the music—music again, always the music; not a cloak for the libretto, but its skin—that defines the piece, and in *Kityezh,* as in *Tristan* and the *Ring,* the musical vocabulary carries the plot action beyond the borders of literalism. Grasping primitive imagism, Rimsky refashions the ring of the tocsin as the ring of insanity in the coward's mind in his mad scene; earlier, the pealing of bells signals a mist that enshrouds the Russian safehold for protection, thus endowing the forces of nature with psychological and religious valences. The permutations of Wagnerian leit-motif, in masterful hands, can narrate the stuff of legend down into legend's depths, and, not surprisingly, such depths are invariably spiritual. Richard Wagner keyed the sacred assumptions of the Camerata—the ceremonial—with "music of the future" put to lore of the past; so, then, did Rimsky respond with a trinity of Kityezshes, Smaller Kityezh, Greater Kityezh (the invisible city), and, in an apotheosis, Celestial Kityezh, an Elysium of perfect beings.

For all their removal from the mainstream of Western art, the Slavic peoples have no monopoly on art as national expression. But there is something to be said for the pure, almost retrogressive ethnicity of Russian opera, nowhere more splendrously expressed than in this holy trio of Kityezhes, with its metaphysical embrace of the green world and its exhilarating mystery-play panoply of battles against the heathen, nick-of-time miracles, Slavic church melodies, and divine transformation. One part ritual spectacle, one part tender love story, and one part unkempt cutup, *Kityezh* is not so far descended from the old story-ceremonial as all that. We are not as fluent in our mythic language as our forebears were, for they knew the stories—we have to look them up. Yet we respond.

However. The early twentieth century was no heyday of mythic opera anywhere in Europe—not even in Russia—and from the sacred, Rimsky, to the satiric. *Zolotoy Pyetushok* (*The Golden Cock*, 1909), to another libretto by Byelsky, dropped hints of the Tsarist fumbling in the Russo-Japanese War into Pushkin's opaque fairytale poem, this time with music that distances rather than envelops the audience. As in *Kashchey*, the supernatural world moves in oriental chromatics, the mortals in plainer chant, but both camps were scored for spoof as deftly as

Kityezh was for credence. Rimsky had expected to make his farewell to art with the Russian *Parsifal,* as Wagner had done with the German, and it was while completing *Kityezh* that he completed his memoirs; but as it happened he closed his account books with an "artificial" puppet comedy. A startling finish for a man so enticed as Rimsky was by folk legend, isn't it?—unless one considers that he was one of the few real geniuses of the late nineteenth century, and thus was bound to cross over from one era to the next. Known to cliché as a brilliant orchestrator and revisor of other men's work, he was much more: music dramatist, folk melodist, and integrator of forms.

Witness *The Golden Cock,* a rare enough bird, to be sure. Its prize puppets are foolish King Dodon and the Queen of She-makha, as caustic a siren as the story hour has ever dispensed. An Astrologer frames the action in a prologue and epilogue; the puppeteer of the show, he also takes part in it, presenting Dodon with a magic cock who functions as an early warning system, crowing repose in peace and havoc in wartime. In a cute conceit that stands midway between *Ring*-cycle leitmotif and dode-caphony, the cock's call to arms is sung to an inversion of the all-clear, varied at one note by a half-tone that marks the differ-ence between restful and spine-scratching.

Dodon meets the Queen of Shemakha while surveying the outcome of a disastrous battle in which, following his own cam-paign strategy, one half of his army has killed the other half. To music of lazy luxury, the Queen appears from nowhere, prom-ises to conquer Dodon, flirts with and ridicules him, and then agrees for no apparent reason to be his wife. When they arrive at Dodon's palace, the Astrologer materializes and demands the Queen in payment for his cock, whereupon Dodon slays the ma-gician with a blow of his sceptre. This vastly amuses the Queen,

but the cock, shrieking its danger signal one last time, darts at
the king and kills him in turn with a peck of his beak. Both
Queen and cock promptly vanish, leaving Dodon's people to
mourn their ruler, and leaving the audience entertained but un-
moved. Yet here comes the Astrologer again, before the curtain
as at the start of the evening, to tell the house that only he and
the Queen were living beings. "The rest?" he explains:
". . . delirium, a dream, a pale phantom, emptiness. . . ."

As in another farewell to art, *Die Zauberflöte*, the authors do
not scruple to make themselves entirely clear. We are told that
Mozart and Schikaneder's Queen of the Night might well be
Maria Theresa and Sarastro's secret society all Freemasonry, and
so might Rimsky and Byelsky's Dodon be Nicholas II and the
chromatic strangers deputies of insidious Japan. More insidi-
ously, posit Dodon as Nicholas but read the Queen as the unapproachable (and foreign) tsaritsa Alexandra, unwittingly in
league with the court "magician" Rasputin, who won his influence by apparently protecting the hemophiliac tsaryevich with
prayer. The timing would be right, for Rasputin first visited the
royal family in 1905, four years before *The Golden Cock*, so this
is all very well as far as it goes, which isn't very far. But one
thing about the opera *is* unambiguous—its tone, which holds in
every measure to the tenets of the satirical smirk. For all its
sinuous beauty, the Shemakhan Queen's vocal line is more sinuous than beautiful, leering and connivant, and the court of King
Dodon is disclosed to music of the most outrageous pomposity.
Even in "tragedy," when Dodon keens over the bodies of his two
sons, slain on the battlefield, Rimsky lets him wail in outlandish
sequential intervals, reversing the tropes of romantic opera (including the lullaby scene and the triumphal march) with grotesquerie and a wink at the public.

This reversal amounts to the three-act equivalent of the banana-peel overturn, and we speak, then . . . now . . . of epoch, for some periods are stronger in comic perpetration than others, and this twentieth century will revel in it. The loving grace of Wagner's *Die Meistersinger* and Verdi's *Falstaff*, with their grand humanistic finales, will be thought counterproductive as of *The Waste Land* and *Vile Bodies*. In the romantic era, such outlanders as Dostoyefsky's devils and Büchner's rag dolls hugged the outskirts while Manfred and Faust descried the demons. But in the twentieth century the demons move in with the perpendicular conjunctions of comedy, and their participation in the theatre ritual calls for music less moving than it is jibing, for demons will have their caprice.

But the romantic era had not yet done itself in, and Arnold Schoenberg's symphonic song-cycle *Gurrelieder* (*Songs of Gurre*), a dramatic oratorio devolved on Wagnerian principles, came to light as late as 1913 (though composed in 1900-01). In its gigantic orchestra (ten horns), its cyclic leitmotifs, its nature imagery, and its Flying Dutchmanesque story, *Gurrelieder* administered the Wagnerian legacy with an amazing melodic charity, ratifying the position of the nonlinear musical narrative as it had come down from Berlioz' *La Damnation de Faust* and would progress on to such works as Debussy's *Le Martyre de Saint-Sébastien* and Stravinsky's *Perséphone,* all of them turning on a mythic premise—all, to be exact, dealing with heroic redemption. In *Gurrelieder,* Schoenberg even arranged for his tapestry of solos with chorus to open and close with depictions of nature play; as Wagner commenced the *Ring* tetralogy with an impressionist stylization of the "creation" of water turning gradually into the sempiternal flow of the Rhine via tuneless E Flat Major scales, *Gurrelieder* begins (also in E Flat) with a timeless sunrise, closing with a transformation of the same music, now in the primevally basic key of C Major, the *diabolus in musica* of the modern age. But just as *Le Martyre* and *Perséphone* employ spoken dialogue, Schoenberg ripped at the romantic canvas with one scene in *Sprechstimme,* the musical inflections of normal speech that were soon to become the *lingua franca* of expressionist opera, the human agony *in secco.*

Gurrelieder, of course, is not an opera, and the fact that it could be staged isn't the point. (Anything can be staged, if one is determined to stage it: ballet companies lay plots on symphonic form and tread upon Haydn and Chaikofsky as a matter of course.) But Schoenberg's contribution to musical dramatization is very much the point, whether in one-act monodrama, chamber song-cycle, or in this mammoth concatenation of ballad epic, Gothic tableau, and word-dance of death. At the height of the terror, when the hero gallops into the sky with his vassals—ghosts, the lot of them—Schoenberg's Speaker lures the fantasy into deformity in dialogue metred and pitched, the Word sliding around in untidy music: "Mister Goosefoot, Mrs. Gooseplant, quickly hide yourselves now, for here comes the wild hunt of the summer wind!" Meaningful, full of the "wrong" meaning, up for antagonism, the satiric adjunct of opera takes the stage in the 1900s, and the Word will shape the artwork and its expressive materials. Here comes the wild hunt of transformation, for even amidst the romantic apparatus of *Gurrelieder,* Schoenberg grasped at and held the fulcrum of era, confirming the obsolescence of the melodramas and historical pageants that had captured, and were to some extent still in charge of, the stage.

Striking a new mold, then, neoclassicists will emerge with a portfolio of the old devices, the escort of the profane dramatic vocabulary—short works (as in the intermezzi of the early Ital-

ian comic operas), either spoken or half-sung, half-spoken dialogue (a twist on the old *opéra comique*), a reprisal of comic
invention—and the music will take on a new tone, a tone more
often perverted than affecting, less a sigh than a snigger. Most
importantly, the tireless legends of antiquity will prove as fit for
retelling as they did when the Camerata "revived" Greek tragedy
in the sixteenth century.

This Flying Dutchman theme, for example—the wandering
sinner in search of salvation—which Wagner used in *Tannhäuser* and *Parsifal* as well as in *Der Fliegende Holländer*, was re-
adapted by Béla Bartók and his librettist, Béla Bálazs, in the
one-act *A Kékszakállú Herceg Vára* (*Duke Bluebeard's Castle*),
merging Western tradition's choice uxoricide with that quintes-
sence of romantic wish fulfillment, the redemptive woman. Bartók's Judit encounters the locked doors that greeted Dukas' Ariane, but here one hears music of terrific scope, for Bálazs opens
these Magyar portals on the seven ages of man: a torture chamber (the terrors of youth), an armory (protected maturity), a
treasure hoard (success), a garden (leisure), the horizon
(power), a lake of tears (the sorrows of age), and, as in Maeterlinck, the forbidden seventh door, which eternal woman *must*
have opened to her. Blood has been seen everywhere, and Judit
suspects that the seven doors hide the corpses of Bluebeard's
former wives. Sadly, Bluebeard hands her the key, and as she
opens the seventh door, the fifth and sixth slowly close, disen-
titling Bluebeard of his passage to death and peace. His search
will perforce continue, for the faithless Judit has failed him, as
did his first three wives, who now appear, and whom Judit joins
in darkness behind the seventh door.

Bálazs' libretto offers a windfall of tropes—the curious woman,
the haunted manor, the doors of secret bounty, the theologically

charged number seven—all enmeshed in the dark idyll of the lonely voyager. Truth to tell, its distillation of classical technique and "objectivity" notwithstanding, this twentieth century will never succeed in burying the goblins of the nineteenth century; whether in subject matter, scoring, vocal thrust, or formal design, the patterns set down by the giants of the earlier century are always with us. Musically, too, *Duke Bluebeard's Castle* recalls Wagner's thunderous footfalls, but in the transparent imagery of impressionism, and like Debussy in his sole opera, Bartók let the cadence of his native tongue guide the musical material.

Hungarian is not an Indo-European language, and its rhythm, even when sung, falls quaintly on Western ears. Bálazs penned *Bluebeard* in trochaic tetrameter, the "Ancient Eight" of Hungarian folklore but a trumpet of doggerel in English ("Double, double toil and trouble, Fire burn and cauldron bubble," not to mention *Hiawatha*). No matter, though, for Bartók endowed the script with a highly textured score, monumental—the Maeterlinck operas redone as epics, with a burrowing deeper in, farther back, in time.

Bartók composed his opera in 1911, but it wasn't heard until 1918, having been decreed unperformable (yes; so were *Fidelio* and *Tristan,* or so the singers said), and even today it garners repute far more than performances. That will prove a leitmotif of modern opera, that uneasy Respect, as recurrent as any Wagnerian reference, that fear of the unworkable masterwork, economically prohibitive, rough on the vocal chords, and inimical to lyric opera buffs everywhere. But the quest of contemporary genius is remorseless, and no amount of professional, critical, and public apathy seems to stop composers and librettists from renewing the sacred and profane with their bizarre new models, though the sacred grows increasingly abstruse, the profane increasingly brazen, and the two in combination ignoble.

Some composers had it easier than others. A man such as Richard Strauss made himself ripe for respect with his early blockbusters—*Salome, Elektra,* and *Der Rosenkavalier*—insuring his trickier later works occasional revival and reassessment. This *Elektra* was at first deemed unperformable for its "ugly" vocal line and monstrous orchestration, most notably by Ernestine Schumann-Heink, who created the role of Klytämnestra at the Dresden premiere in 1909, and who forever after swore that at one rehearsal she could hear Strauss egging on the conductor, Ernst von Schuch: "Louder, louder the orchestra! I can still hear Frau Heink!" Heink survived without *Elektra,* and *Elektra* without Heink, but *Der Rosenkavalier's* immediate successor, *Ariadne auf Naxos,* had much greater difficulty in establishing itself, for it tangled rather heavily with the problem of formal profile. Remember the pseudomythic parade of baroque opera and the ballad comedies of the fairgrounds, for here in *Ariadne auf Naxos* is an early junction point of their two arts, the poetry with the logic, the *sensuel* with the *gai primitif.* Other, perhaps better-loved, Strauss operas await us in a later chapter; just for now we seek the exceptions to the rules.

Its libretto, like those of *Elektra* and *Der Rosenkavalier,* from the hand of Hugo von Hofmannsthal, *Ariadne* is less popular than they but remains Strauss and Hofmannsthal's most provocative work, for all the intensity of *Elektra* and the taking charm of *Rosenkavalier*—and this despite the fact that *Ariadne* as known today is nothing but a compromise, an entailment on an original and architecturally catholic idea that somehow just never came off. The intention was to drop an "Ariadne Abandoned" type of *opera seria* into the framework of Molière's comedy *Le Bourgeois Gentilhomme, Ariadne* being the piece performed at

the end of the evening for the upstart Monsieur Jourdain and
his guests. With much of the text cut, this still meant a long
evening, and the premiere in Stuttgart in 1912 did not go over
well, despite an impressive congregation on the production team:
Strauss himself conducted, Max Reinhardt directed, and both
singing and acting principals were imported from Berlin, Dresden, and Vienna, the singers including such experts in Straussian
line as Maria Jeritza, Margarethe Siems, and Hermann Jadlow-
ker. Besides the Ariadne opera, Strauss had composed eighty
minutes' worth of song and incidental music for Molière, and the
King of Württemberg elected to hold elaborate receptions during
the two intermissions, making a long evening insufferably longer.
The hoped-for magic of this theatre-cum-music-theatre confection did not obtain, and the diagnosis for its failure, subscribed to
by Strauss himself, was that the theatre crowd loathed the opera,
and the opera buffs abhorred the play.

Unlikely, that. The cultural public is hardly so stratified, especially in Europe, where art is art until the totalitarians take
over; the people who attend to Schiller, Racine, and Shakespeare
take in Beethoven and Berlioz as well, regrouping at each attraction into what André Gide has termed the fourth unity, that of
the audience. No, more plausibly, this Strauss-Hofmannsthal-
Molière-Reinhardt opus failed because it was (1) too damn long,
and (2) simply ahead of its time, for the rebalancing of words
and music wasn't yet on the docket in 1912. Incidental music
was a given in drama, but full-throated opera was not; furthermore, this *opera seria* was actually two different pieces played
cheek by jowl, the *Ariadne* proper and a cheesy commedia, for a
fireworks display is on tap and Jourdain decides that his two
rented diversions, the tragedy and the farce, must be performed
not one after the other but simultaneously.

Talk about one's sacred and profane: on one hand, the chaste

Ariadne, homeless and alone, waiting for death, and on the
other, the confraternity of comics, prancing and prattling, led
by the arch-coquette Zerbinetta. An amusing juxtaposition, it
needed genius to make it meaningful; genius it received. To
Hofmannsthal, it was a bilateral descent into the faith and fickle-
ness of womanhood, to Strauss an opera like any other, coloratura showpiece, gran duo, and all, but this is the most inspired
partnership in opera's history, and of their six collaborations,
Ariadne is arguably the masterpiece. It might have languished
in the attic of the dispossessed, along with the elephant's-foot
umbrella stands and the hip baths, had not the two men given
up on Molière after the premiere performances and contrived
a prologue to precede the opera on its own, set in the home of
"the richest man in Vienna," where a fireworks show again
forces the singers and comedians to cohabit. The music of the
prologue quotes bits from the opera, but it is conceived of a
wholly different métier, much of it given over to light recitative,
one of its characters dealing solely in speech, snatches of song
bursting out here and there—a singing play.

The prologue follows the last half hour before the performance, during which the two casts, the clowns led by a Dancing
Master and the *artistes* by a Music Teacher, prepare their
makeup, bicker, and have tantrums. Hofmannsthal added one
new principal, the Composer of the opera, a combination of Mozart and Cherubino. Too young to be a baritone and too important to be entrusted to a tenor—the entire race of whom Strauss
held in contempt—the Composer is a trouser role, and as he takes
his calling very seriously, he is stricken to learn that his vision
of woman in the ideal is to share the stage with Zerbinetta and
her bad company. "Dances and warbling?" he cries. "Insolent behavior and double meanings?" But for the intervention of the

charming Zerbinetta in what is either a rare moment of compassion or the millionth installment of her flirtation serial, the Composer is ready to start the *querelle des bouffons* all over again. Raging at the crassness of his patron and the patron's guests, he condemns the lack of soul that calls for high-jinks to undercut nobility: "These infinitely vulgar people want to build bridges from my world over into theirs!"

But of course! That's exactly the point, and it works superbly because the two worlds do ultimately connect—six of the sublime, half a dozen of the ridiculous. In the opera, Strauss tenders transport to the classical characters and folderol to the harlequins, but as Zerbinetta points out, women are all alike, and she and Ariadne are at the same crossroads, waiting for the next god to transform them. Zerbinetta has just run through a rota of lovers, and Ariadne has been left on Naxos by Theseus: they're both between engagements. True, Ariadne believes that her next lover will be death, but it is Bacchus (Dionysus) who arrives, as per legend—not the Dionysus of Euripides' *The Bacchae,* the devouring god who levels Thebes for resisting him, nor yet the happy swillpot hymned in the Pedrillo-Osmin duet of Mozart's *Die Entführung aus dem Serail,* but a big, loud tenor in vine leaves who carries Ariadne to the stars. As they stand enraptured, heroine and deity, woman and man, Zerbinetta pokes her head in and sums up the case for her sex: "When a new god comes," she explains, "we are given away . . . befuddled."

A pivotal work in its new-fashioned word-music ambience in the prologue and in its fluent conjunction of the sacred and the profane in the opera, *Ariadne auf Naxos* was a tour de force for Hofmannsthal as much as for Strauss, as his libretto showed itself richer in poetry and wit than anything that had passed before. A radiant humanism shines through the piece at every turn, whether in the Composer's ingenuous self-dramatization, Zerbinetta's practical philosophy, or Ariadne's idealism. To Hofmannsthal, clearly, myth and romance and music and comedy are all fellow travelers, and perhaps no one but he could have guided the trip so surely. How generous the classicism, yet how unforced, and how breezily natural the comedy, as in the C Major "ariette" of the Dancing Master in the prologue, when Zerbinetta, thinking that her act is to succeed rather than join the opera, complains of the difficulty of rousing an audience numbed by *opera seria*. Her colleague reassures her:

> Im Gegenteil. Man kommt vom Tisch, man ist beschwert und wenig aufgelegt, man macht unbemerkt ein Schläfchen, klatscht dann aus Höflichkeit und um sich wach zu machen. Indessen ist man ganz munter geworden. "Was kommt jetzt?" sagt man sich. *Die Ungetreue Zerbinetta und ihre Vier Liebhaber—*ein heitres Nachspiel mit Tänzen, leichte, gefäll'ge Melodien, ja! eine Handlung klar wie der Tag, da weiss man, woran man ist. "Das ist unser Fall," sagt man sich, da wacht man auf, da ist man bei der Sache! Und wenn sie in ihren Karossen sitzen, wissen sie überhaupt nichts mehr, als dass sie die unvergleichliche Zerbinetta haben tanzen seh'n!

> (On the contrary. They'll be coming from the table heavy with eating and sleepy with drink, and in the dark they'll doze through the music unnoticed, and clap out of courtliness and try to look reverent. Meanwhile, they're wide awake and feeling merry. "What comes next?" they ask. *The Fickle Zerbinetta and her Four Lovers—*a merry afterpiece with dancing, simple tunes, and a premise clear as day . . . just what the doctor ordered. "Ah," they say, "this is more like it." They're all ears and ready for art. And when they ride home in their carriages, they'll remember only that they saw the delightful Zerbinetta at her sport!)

As for Strauss, he had got sentiment down wonderfully in *Der Rosenkavalier,* but neither in that work nor in his earlier comedy, *Feuersnot,* did he capture the heart and the mind of satire as well as he did here. The clowns' quintet, "Es gilt, ob Tanzen," carries jolly insincerity just so, but one is surprised by the serene beauty of its coda, "D'rum lasst das Tanzen"—and when did any bel canto canary exhibit the life-true conviction of Zerbinetta's display piece, "Grossmächtige Prinzessin," with its vibrant, hedonistic *cogito* and dazzling rondo, as flighty as sky-caressing coloratura knows how to be? From its resurrection in Vienna in 1916 (with Jeritza again, and Lotte Lehmann as the most urgent and impressionable of Composers), *Ariadne auf Naxos* has lived a double life, as "difficult" elitist fare and—in German-speaking countries—as repertory item.

Well, if legend and commedia can so resourcefully coincide, it remains only for someone to strike out for the last Indies and beach his craft on the shores of profaneness, moving into satire and the Word. Not that this *had* to happen, of course—simply that it was going to, and did, for the supermusical romantics had lost the will to renew, except in Italy. Moreover, the pseudohistorical settings and characters that had predominated in bel canto and grand opera had run their course, and were ceding pride of place to the more significantly universal images of legend. Oh yes, myth always proves adaptable, from era to era, but often enough it is not a question of the myth of transformation, as in *Les Troyens* and the *Ring,* but rather the cagier, acid myths of comic stability—the perpetuum mobile of Zerbinetta, say, as opposed to the cool transcending of Ariadne.

Well, let us try *these* mythic tropes on for size: the three riddles, the nameless stranger in town, the virgin princess waiting to be won. We've seen them around often enough; so centroidal

a figure as the *Nibelungenlied* Brynhild imposes a tripartite test
on her suitors, though in the form of track-and-field events rather
than riddles. Count Carlo Gozzi wrapped up the riddles, the
stranger, and the princess for Venice in 1762 in his "fiaba chinese teatrale tragicomica," *Turandot*, set in antique Peking,
where an unknown prince solves three riddles (answers: the
sun, the year, and, appropriately if out of timesync, the "Lion of
the Adriatic" [Venice]) to conquer the untouchable Turandot,
who has fallen for him at first sight but must fight back for the
sake of the imperial pride. Humiliated by defeat, Turandot
threatens to stab herself at the altar, so the prince gives her a
sporting chance: if she can guess his name, she can stay a spin-
ster. Turandot's confidential slave, who knew and loved the
prince in another country, reveals his identity to gain her freedom, but ultimately Turandot accepts the prince as her husband
anyway. It's a protean tale, capable of any posture. It can move,
or frighten, or cheer, or teach. How shall it go?

Gozzi wrote to reclaim the Venetian commedia heritage, at
that time in desuetude because Carlo Goldoni had purged the
stage of vernacular improvisation and fantastical doings—remember it was he who wrote the libretto to Piccinni's circumspect
little charmer, *La Buona Figluola*. But remember, too, that it was
Venetian opera that first blended comic incongruities with seria
attitudes, and as part of his drive to restore the freewheeling
mummery of San Marco, Gozzi larded his shows, whatever their
settings, with the eternal masks—the stutterer, the cuckold, the
blowhard, the shrew—all the loiterers and hustlers of the comic
demimonde. For *Turandot* he proposed Pantalone, Tartaglia,
Brighella, and Truffaldino, not as extracurricular novelties, but
to set the tone, to lock the arbitrary circuits of fairytaledom into
the pragmatic rationalism of the everyday. (Besides, he had four

comedians at his disposal in the company.) Within such a framework, fantasy disrobes its magic and leans to grotesquerie: this is
Turandot.

Even before Gozzi, the plot had served an opera by one D'Or-
neval, *La Princesse de Chine* (1729), and it surfaced again in
Giacomo Puccini's last work, but a close adaptation of Gozzi first
served the Italian-German philosopher-king and sometime composer, Ferruccio Busoni. Like Maeterlinck, he was attracted to
the remote whimsey of the puppet play, and scored the incidental
accompaniment to Max Reinhardt's production of the Gozzi ex-
travaganza in 1911. The subject proved so distracting to him
that he expanded his suite into a shortish, two-act opera, writing
both words and music and following closely in Gozzi's shadow
(he did rewrite the three riddles, however, which now answered,
epistemically, as human reason, mores, and the arts).

Busoni's *Turandot* (1917) bowed as the antithesis of romantic opera as surely as Busoni himself faced down the generations
to consummate the implications of Bach's keyboard in harmonic
and structural philosophy—like Coleridge, in Byron's phrase,
"explaining metaphysics to the nation." Though Busoni might
stray from pattern, as in a partly romantic, five-movement piano
concerto (with male chorus) of Herculean demonism, his stage
works heralded the neoclassical revival that was to regenerate
European musical circles.

Of Busoni's four operas, *Turandot* is by no means the best, but
it is perhaps the most exemplary. Reduction of forces is a key
here—small forms, with arias so tiny as to be hardly there at all;
fleet pacing, this scene, that scene, the next, like dances at a
ball; simple part writing for the chorus and concise, almost cadav-
erous dialogue between musical numbers . . . a grand opera in
miniature. Unlike Puccini's *Turandot,* Busoni's grants nothing

to the spectator in the way of sentimental zap or *le merveilleux;*
on the contrary, he keeps his public at a distance throughout,
showing them a funhouse mirror reflection of romantic opera.
The profane is heir to the sacred.

Gozzi's beloved masks are in their element here (what, Truffaldino, Pantalone, and such in legendary Peking? Yes, yes indeed) and the entire score adapts to their note, hacking away at
the author-audience communion of the sacred play with materials not unlike those that came in so handy for Rimsky-Korsakof when he wrote *The Golden Cock.* Empty-headed, thinly orchestrated Chinese pastiche destroys any illusion of place,
especially in the rather bouncy funeral march in honor of Turandot's latest victim (for to fail in the riddles naturally means
death), and in the riddle scene a piccolo run ascending into its
shrillest limits dispel any mystery that the impressionable operagoer may have spun for himself.

Now. For all its rawboned lamenting, spiraling outward in
aseptically concentric circles from the heart of *The Waste Land,*
this is a rich age. The time: 1900 and onward; the place: Western music drama; the premise: that romanticism will never die,
though at times it may seem somewhat posthumous. The pivots
explored in this chapter reveal that new art will be built on old,
and new expression be made of old action, whether as naturalism
or myth, as profane satire or sacred musicality. These are but the
pivots; much more is to come, in so many varieties that every
slogan, every label, will take its exceptional corollary.

As Busoni's Unknown Prince puts it, in his fifty-second entrance aria after a fifty-five-second prelude, "I sense the extraordinary."

III

*The Last Days
of Romance*

I. FRANCE

The transformation from the nineteenth century into the shy
new world in France was most securely negotiated by Debussy
and Dukas in their one opera each, but to do it justice, the *tragédie lyrique* of the early 1900s is best identified with the playing
out of the Wagnerian craze and, on other stages, the declining
powers of the aging Jules Massenet. All of the above shared a
zest for anti-naturalistic subject matter—*Louise* was but an instant, a bit of grout come loose in the wall—as the fantastic and
historical settings of the 1900s were rejoined by classical motifs,
with a Pénélope here, an Agamemnon there, and often if not
everywhere an Aphrodite.

Paris was where Richard Wagner suffered his first international disaster, with the famous disruption of the second and
third performances of *Tannhäuser* at the Opéra in 1861, when
the Jockey Club and various standpatters gave an opera of their
own in the boxes and stalls. Still, it was Paris that gave *la maison
Wagner* its finest continuity for two decades after the founder
died in 1883. The most meticulous Wagnerian model was run
off by August Bungert in his *Homerische Welt,* but a Frenchman, Ernest Reyer, wins the palm for the Best Wagnerian Entry
Not by Wagner in his *Sigurd* (1884), a look at the *Nibelungenlied* à la Française, not excluding some Valkyries in the corps de
ballet.

The French were pioneers in the realm of program music, en-

couraged as they were by Liszt, and if Gluck had already forced
the issue of a programatic orchestra in opera, Wagner clinched it
so tellingly that for many a while many composers assumed that
only the mythic reaches of Teutonic antiquity would do for a
libretto: if it didn't have dwarves and nixies in it, it wasn't an
opera. Speaking in broader terms, the art of the 1890s and early
1900s was dominated by *Jugendstil* and symbolism if not whole-
hearted myth, and a kind of remetabolized romanticism outdrew
everything else until the neoclassical revival intruded in France
and diatonic jazz surfaced in Germany—only the Italians, of the
international companies, held onto tradition.

But in 1900 the lure of Wagner was as strong as ever, so
Vincent D'Indy could allude rather heavily to *Parsifal* in his
Fervaal (1897) and Ernest Chausson go the *Tristan* route in *Le
Roi Arthus* (King Arthur, 1903). Even Ernest Guiraud caught
the bug in his *Brunhilde,* retitled, on second thought, *Frédégonde*
(1895, finished by Saint-Saëns), though he couldn't resist setting it in Paris. How could they know that this was the twilight
of those particular gods, with stage works by Maurice Ravel,
Gabriel Fauré, and Eric Satie on the horizon? In the last hours
of the French Wagnerian cult, D'Indy produced *L'Etranger*
(*The Stranger,* 1903), using the Flying Dutchman tale in almost exactly the same way Gilbert Bécaud was to in his *Opéra
d'Aran* in 1962. D'Indy's Stranger was born to be emblazoned on
the art-nouveau posters of the day, a mysterious and somewhat
Kabalistic wanderer who upsets the routine of a French fishing
hamlet in his search for absolution by some womanly life force,
here named Vita. *L'Etranger* was perhaps the last stand of *le
Wagnerisme,* though the German wing carried on for a bit;
for Paris remained the closest thing in Europe to a central dis-
pensary of art, and it was in Paris that the unceasing renewal of

form had already crossed the bar in the pointillistic score to *Pelléas*.

Still, nothing happens all at once, and the acknowledged grand master of French music drama until his death in 1912 was one who had won his spurs way back in the 1870s, when Meyerbeer's corpse was still warm and Wagner in his prime. Sometimes dubbed Mademoiselle Wagner for his alleged Teutonic practice and mauve autograph, Jules Massenet occupied the post left vacant with the passing of Gounod, Thomas, and Delibes. Bizet subverted the bourgeois opera of sentiment in the naturalistic *opéra comique Carmen* in 1875, but Massenet kept it alive in a winning streak from *Hérodiade* in 1881 through *Manon, Le Cid, Werther,* and *Thaïs* to *Cendrillon* in 1899.

The charges of Wagnerism and effeminacy in Massenet are unfounded. He did conceive a succession of vehicles for a succession of odalisques, it's true, and *le voluptueux* pervades, but Massenet was as trusty with heft as he was with grace, and as for the shadow of Bayreuth, it darkened his Elysium but seldom. For one thing, Massenet worked in closed forms until his very last operas, and was more oriented to spoken dialogue than to symphonic scoring. He evolved a wonderful style of recitative *très musical,* usually in 6/8 or 9/8, conflating melody with the effortless rhetoric of spoken French, but he resisted the cyclic leitmotif, preferring his own notorious *phrase massenétique décadente,* a Big Tune to be repeated rather than modified at high-mettled moments.

Like Rimsky-Korsakof, Massenet worked his way through genre. By the time 1900 rolled around he had made a go of most of the types of the time—oriental fantasy in *Le Roi de Lahore,* grand opera in *Le Cid,* a touch of Wagnerian love-dream in *Esclarmonde* (complete with quotations of the falling semitone,

from the *Ring,* that used to be known as the Woe Motive), verismo in *La Navarraise.* But when the century turned, Massenet's
powers were in abeyance, though he was still the reigning
monarch of the art. Each Massenet premiere brought the music
world out in force, more often than not to Monte Carlo, where
the composer had an empathetic relationship with the intendant,
Raoul Gunsbourg. There was a touch of show business in all
this, not because Massenet set such store by production values, which, however, he did, but because he more than anyone
of his time subscribed to the star system, subclass raving diva.
His habit was to design a lead role for a particular singer's talents,
coaching her, inspiring her, and where possible making her his
mistress. He did seem to have an understanding with sopranos
and contraltos. The blonde Californian Sibyl Sanderson emerged
as Esclarmonde and Thaïs, the respectably tempestuous Emma
Calvé got *La Navarraise* and *Sapho,* Massenet's foreign ambas-
sador Mary Garden virtually lived the name part in *Chérubin,*
and the gorgeous Lucy Arbell was handed no less than six roles
by the composer. All of this is just so much bespoke creation, devoted as much to the artist as to art, and without the stars Massenet's works could easily fall flat.

Sadly, by the twentieth century they were starting to droop
even with the stars, what with a parade of almost annual titles
that now are one with Gondwanaland, submerged in oblivion.
Roma, Panurge, Amadis, Cleopâtre . . . where are they now?
Classical and medieval pastiche leaden with recipe, where, indeed, were they then? 1914, the year of *Cleopâtre,* Massenet's
swan song, saw the premiere of Zandonai's invigoratingly dramatic *Francesca da Rimini,* Stravinsky's *Le Rossignol,* of a truly
new sound, and in Massenet's own territory, Henri Rabaud's
Mârouf, Savetier de Caire, which adapted the décolleté lyricism

of bourgeois romance to an edge of modern self-ridicule. *Cleo-pâtre* wasn't even in the running.

For the man who had ravished the world with *Manon,* eclipse would have been all the harder to confront had Massenet to confront it, but as long as he lived his works enjoyed a vogue, and the front-rank casts that the young Verdi warranted as the catalysts of success or failure were invariably Massenet's to command. Mary Garden singlehandedly made a standard of *Le Jongleur de Notre-Dame (The Juggler of Our Lady,* 1902) after Massenet reworked the tenor lead for her, and the role became something of an ad nauseam signet for Garden. Critics demanded an opportunity to gauge the piece as originally composed, but when they got their chance they found themselves almost alone in the theatre, so identified was the work with its chief exponent.

Qua work, *Le Jongleur* is typical Massenet, colorful, brilliantly orchestrated, somewhat repetitive, and lacking in resonance between the climaxes. An adaptation by Maurice Léna of a story by Anatole France, *L'Etui de Nacre,* it pictures a wandering minstrel of the middle ages who is taken in by some monks and discovers a need to render some service to God's glory. One of Massenet's few austere scores, *Le Jongleur* offered but one highlight, an antique ballad about Mary fleeing the Romans with her infant Son. Parables being what they must, a luscious rose refuses to hide the child for fear of spoiling its petals, but a humble sage plant gallantly aids the Madonna. Inspired by the song, the equally humble jester promptly offers homage to a statue of the Virgin employing the only métier he knows, the theatre. The monks are scandalized, but Mary accepts the singing and stepping in a miraculous apotheosis, during which Jean, the juggler, dies, in bliss, of exhaustion. The shortish piece is suffused with medieval noises—the viol, the droning bass, the oboes squealing in their upper register, a modal dance, and "l'eternelle pastourelle de Robin et Marion," performed for the Virgin—but though the moral cellophane of the story is deeply felt, there is nothing profound or even essential in it. Garden's reputation notwithstanding, trouser roles are now in quite welcome disrepute, and *Le Jongleur's* occasional recent revivals have restored the lead to the tenor clef.

But if Massenet lacked the breadth for the truly sacred, he sure had the swank for the profane, especially with the help of a vivacious leading lady and his usual affinity for the eclectic flotsam of pastiche. Nowhere nearly so successful as *Le Jongleur*, but easily Massenet's best work in the twentieth century, was *Chérubin* (1905), detailing the further exploits of Beaumarchais' little page, "Cherubin d'amore." A Monte Carlo special, the *Chérubin* premiere disclosed three star attractions—Mary Garden *en travesti* in the title part, the insanely beautiful Lina Cavalieri as the dancer L'Ensoleillad, and Marguerite Carré in a role "pour soprano de sentiment," that of Nina, confidante and secret admirer and, at last, fiancée of Chérubin.

The opera was not based on anything by Beaumarchais, actually, but rather on a play by Francis de Croisset that was withdrawn after a rickety rehearsal period at the Comédie-Française. Massenet saw its potential as a "comédie chantée," a reprisal of the quasi-*opéra comique* style of *Manon* retaining the savor of the *phrase massenétique* but asserting an independent verbal air. Henri Cain helped de Croisset revamp his rejected script into a libretto in which Garden as Chérubin would fall in and out of love, fight duels, throw parties, and be generally full of the dickens. The rapscallion adolescent is seventeen in this installment, guided by his teacher, Le Philosophe, harried by a Count, a Duke, and a Baron, flirted with by the wives of the last two, picked up and then tossed aside by the dazzling Ensoleillad, and adored throughout by the patient little Nina.

The set pieces that intrude on the theatre of the sublime are quite applicable for the ridiculous, and such a comedy as *Chérubin* can survive a hint of the vaudeville. Act I offered a *fête pastourale* that Lully would not have been ashamed to conduct, Act II a husky Spanish dance, the Manola (tripped by Cavalieri), Act III a mandolin, guitar, and flute aubade, and the whole evening was prefaced with a scintillant rondo overture.

Chérubin is certainly Massenet's forgotten trump card, played when the deal was beginning to run against him, but for all its charms, its mercurial debate between song and speech, it is perfectly characteristic of the old Massenet and no breakthrough into the real 1905, time present. Subtly paced, tender, worldly, luxurious, farcical, it still couldn't resist the sympathetic engraving. At the opening of the final act, with three duels only hours away, Chérubin is glimpsed making out his will to music vulnerably delicate and yet jaunty, and the verses hit the mark:

> *Si je reçois un coup de dague,*
> *Si ce soir je dois trépasser,*
> *A Nina je donne ma bague . . .*
> *Pour être un peu son fiancé;*
> *A L'Ensoleillad rose et brune,*
> *Dont l'amour un soir m'a grisé,*
> *Je donne toute ma fortune . . . .*
> *Et c'est bien peu pour son baiser.*

> If I should receive a dagger wound,
> If I must die tonight,
> To Nina I leave my ring . . .
> For having been her sort-of fiancé;
> To L'Ensoleillad the rosy brunette,

Whose love one night intoxicated me,
I leave my entire fortune . . .
It's little enough to pay for her kiss.

And yet, withal, there was a touch of sardonic commentary
from the composer, and about time, too: in the final measures of
the opera, when Chérubin takes Nina as his love, Le Philosophe
begs him to dispose of his souvenirs d'amour, the latest being the
Countess' ribbon. This the boy refuses to do, and as the orchestra gently quotes the serenade from *Don Giovanni*, an onlooker
comments, "He's Don Juan!" And Le Philosophe, gazing sadly
at Nina, answers, "And she's Elvira."

After *Chérubin*, Massenet's trip was a downhill slide, even if
the premieres were attended with reasonably bated breaths as
long as Lucy Arbell, Lucienne Breval, and Fyodor Shalyapin
were willing to create the leads. As an earnest of his share in the
revolt against naturalism, Massenet delved into classical myth,
but the texts called for music of a more intrepid charisma than
that of Massenet, as always more sensual than profound. On a
pair of librettos by Catulle Mendès, Massenet covered the Ariane-
Theseus-Bacchus stories in *Ariane* (1906) and *Bacchus* (1909),
splattering the stage with the contraption of legend, including
descents into hell and Dionysiac parties in Nepal, all too feverish
and vainglorious, the kind of French operas that peak in their
ballets.

Massenet did stage a comeback, though, in *Don Quichotte*
(1910), one of his last works and always a sell-out when Shalyapin was available to apply his fanatic plastique to the figure of
Don Quixote. A gleaning of Cervantes in five quite slim acts,
Henri Cain's libretto sought the ecstasy of quest, and while Massenet held to his palette of heterogeneous pastiche, he failed to
capture the double image of *noblesse* and *canaille* in the Don

and his sidekick Sancho that Richard Strauss had so aptly presented in the cello and viola parts of his tone poem *Don Quixote*. The understated final scene, wherein Quichotte dies stretched out before the despairing Sancho while the voice of the "fair damsel" Dulcinée echoes in the air, was admired in its day: "Mon bon, mon gros Sancho," cries Quichotte, bequeathing his fat little shadow the most beautiful of all his possessions, "the island of dreams."

Consider, though, that France was to lead the neoclassical rout, leaving both realists and traditionalists—and dreamers, especially—bereft of foundation, their podiums pulled down by a sudden passion for austerity in short forms ruthlessly patrolled for symmetrical point. Eric Satie, that enfant terrible of a Nestor, advised and encouraged the next generation before the last was passé enough to resign. Unquestionably, the French are the most analytic of Europe's peoples when it comes to the flux of artistic eras; they shape their perspective and align their options the way some folks order dinner. One has only to recall the *querelle des bouffons* and the Gluck-Piccinni pamphleteering to realize how the climate must have felt to the romantic leftovers who lacked Massenet's flair for casting.

As a matter of fact, Camille Saint-Saëns somehow wangled Nellie Melba for the face that launched a thousand ships in his *Hélène* (1904), at Monte Carlo, though in fact Melba's face couldn't have launched a pincushion. This one-act's worth of Olympian litigation anent the elopement of Helen and Paris (Venus—it should properly have been Aphrodite—seconds the plan with a rash of coloratura; Pallas Athene votes no with contralto pressure) proved insipid and, despite Saint-Saëns' early success with the libidinous character of Dalila, horribly platonic. In any case, Melba was a splendid singer, but an actress of the old school, one eye permanently cast on the blackboard; no, the humdrum French operas of the early 1900s found a more ideal deputy in Mary Garden, who carried them about with her wherever she went. For Oscar Hammerstein's Manhattan Grand Opera and her own company in Chicago she was not only Louise, Thaïs, Mélisande, and Jean, the juggler, but the heroines of *Monna Vanna* and *Gismonda*, written for her by Henri Février after Maeterlinck and Sardou, respectively. Février dealt in diminished chords and tautological motto themes, and though he could compose pleasant love music at the gallop, his operas generally sounded like loutish Massenet; they were attempted, in their tiny currency, only by sopranos such as Garden, or Fanny Heldy, who could spin this straw, for two hours or so, into gold.

The collapse of Paris' hegemony as the world capital of opera has blotted out the once prominent "French style" and deprived international stages of the singers conversant with Gallic practice, so much of lasting interest has perished along with *Hélène* and *Monna Vanna*. Since the 1930s, for example, Dukas' wonderful *Ariane et Barbe-Bleue* has been heard rarely if at all outside of France, and the list of casualties further includes such enticing statistics as Gabriel Fauré's *Prométhée* (1900), a festival work ordered for an outdoor theatre in Béziers. Too loosely adapted from Aeschylus' *Prometheus Bound*, the libretto sagged noticeably, but Fauré responded to the monolithic subject with a monolithic ensemble—four hundred musicians, including one hundred strings trained in from Paris, two separate military bands, eighteen harps, and a chorus of two hundred. This was Fauré's *Gurrelieder*, on one of Western culture's most shattering ideas, but according to those who were there, it was less colossal than loud, the text so weak that a revival at the Lyon Festival in 1970 called for a wholly rewritten libretto.

Fauré again turned to Greek legend in his better-known *Péné-*

lope (1913), a slow-moving but attractive investigation of Odysseus' homecoming. Far from being the silly natural history fluff it is sometimes taken to be, explaining the origin of rocks, trees, whirlpools, and the like, the myth claims a psychological and historical derivation, mostly relating to the primeval changeover from the matriarchal barbarism of queen bee villages to a male-dominated society of warriors and kings. No doubt all this is too far behind us to raise the right hackles in the stalls, but there is yet a mystery to be sought in our mythic avatars—wait, now: the seekers will come—even if the trumpet theme that Fauré assigned to his Ulisse (Odysseus) carries no more weight than the air of neat domestic nobility.

Here is a seeker: Claude Debussy. Still in the grotto of Maeterlinck opera, he contemplated drafting Edgar Allan Poe for the lyric stage, never actually finishing *La Chute de la Maison Usher* (*The Fall of the House of Usher*), but spearheading the drive to a remixing of the forces of word and music. The event was a "miracle play," *Le Martyre de Saint-Sébastien* (*The Martyrdom of Saint Sebastian,* 1911), with a text of dialogue and lyrics by Gabriele d'Annunzio, who was suffering a fetish for the dancer Ida Rubinstein at the time. Nothing that D'Annunzio ever wrote was not distinctly symbolic and sexual, and one Cardinal della Volpe added greatly to first-night expectancy by placing the poet's entire oeuvre on the Index a few weeks before the performance. Some critics obliged the Cardinal by reporting the work to be a blasphemy, despite an unmistakably devotional attitude on the part of both authors and Rubinstein as well; indeed, D'Annunzio took the trouble to compare Sebastian's agony with that of Christ (and Adonis) in a compelling choral passage set by Debussy in a nonharmonic horizontal line bordering on atonality.

Part play, part oratorio, and possibly part untheatrical, this

Mass *en ballet* proved a trial, for while the French could look back on a rock-solid foundation of dialogue, song, and dance, they weren't looking all that hard in 1911, and in the case of *Le Martyre de Saint-Sébastien* they were literally averting their eyes. More to French taste was less rejuvenation of form. They longed for the perky orientalism and bright melodic periods they could bathe in at one sitting. Such a work would be, say, Henri Rabaud's *Mârouf, Savetier de Caire* (*Mârouf, Cobbler of Cairo,* 1914), a sleek Arabian night about an innocent ne'er-do-well who becomes the wealthy son-in-law of a sultan through the agency of an agile tongue and an expedient genie. In Mârouf's graceful "Il est des Musulmans," set to a thrumming ostinato, and in his melismatic "Beauté du cou de ma Saamcheddine," the piece proved as dry as champagne with a prickle of wit to spoof its own fantasy, a parfait set before the bourgeoisie. Not for them the declamatory temperance of Ernest Bloch's *Macbeth* (1910)—they hadn't much cared for Verdi's setting, either, in 1865—not while boulevardiers such as André Messager and Reynaldo Hahn were in town.

In fact, much as the French operagoers attended such exotica as *Mârouf,* as Gabriel Dupont's *Antar* (1921) and Albert Roussel's Hindu opera-ballet *Padmâvatî* (1923)—or even a Gallic *Salomé* (1908) by Antoine Mariotte, set, like Strauss' try, to Oscar Wilde's script—they would never forsake the lighter, word-true stage works of the profane theatre, where Roussel, so fetched by modal India, could experiment with racy comedy in *Le Testament de la Tante Caroline* (*Aunt Caroline's Will,* 1936). The grand opera era put other gods before the Word—spectacle, vocal luxury, and plot intrigue, for starters—but *Pelléas* and *Ariane* reinstituted the integrity of the libretto in their settings of, rather than "adaptations" of, literary sources.

The forte of French musical comedy is its tradition of high-quality composition. Even Gluck, between *Iphigenia*s, penned his *opéras comiques,* so it is no surprise to find Roussel "fitting the tunes" to a libretto by a certain Nino about shady Aunt Caroline's will, which leaves a fortune to the first son born to either of her three nieces—and they've got exactly one year in which to produce or everything goes to the Salvation Army. A lightly scored one-act in dialogue, small numbers, and one choral ensemble, *Tante Caroline* upheld the tenets of the farcical intermezzi of the past, meanwhile giving Roussel a nifty rest-cure from his several exertions on behalf of Greek legend in opera and ballet. As if the condition of Aunt Caroline's bequest were not risqué enough as stated, the efforts of one of the nieces to get issue, never mind how, have their juicy side, and when the heir is at last revealed he turns out to be the byproduct of a forgotten episode in the salad days of another niece, particularly green in judgment. Her aria, "C'était un gars de la Bretagne," relates the touchy business in a delicate barcarolle, for her *gars* was a fisherman and their tryst occurred on his boat. Sly repeated figures and unexpected notes (the melody ends on the sixth of the scale) keep the profane in gear, and Nino's verses are deliciously casual about it all (note the pun on "pèché," meaning both fished [pêché] and sinned):

> *Tendrement il me prit la taille,*
> *Je n'ai pu l'en empecher;*
> *Toute la nuit,*
> *Vaille que vaille,*
> *Ensemble nous avon pèché.*

> He assessed me so tenderly,
> I was hard-pressed to be discreet;
> All that night,
> We cut bait and fished—
> The angling was rather complete.

The attraction of a *Tante Caroline* lies in its clarity, that aura of the *gai primitif* that has fascinated French musicians as far back as the record reveals them, and as recently as next week. Is it perhaps stretching the point to find this grand, outlaw impertinence in an offering such as Louis Varney's *opéra comique, Le Chien du Régiment (The Dog of the Regiment,* 1902), with its French poodle hero, Moustache, its dizzy escapades, its martial exhibit, its tender couplets in which the ingenue so appealingly bids farewell to Moustache when he trots off to war? It is. But *Le Chien* is mere vaudeville; the point is that the machinery of *opéra comique* was being hooked up to more serious endeavors, the sentimental apparatus disconnected but the simplicity, the directness redoubled.

Here is directness: Eric Satie. He, the Prometheus of the generation covered in this chapter, not only passed over the fire to his friends, but hoodwinked the establishment as well, the trickster titan. Was it not he who put a typewriter, sirens, and the sounds of a Morse telegraph key and an express train into the ballet *Parade* that he wrought with Jean Cocteau, of which work Francis Poulenc observed, "The music hall was invading art with a capital A"? Was it not this same Satie who wrote minute, sometimes jarring and sometimes monotonous piano pieces entitled, for example, "Affolements Granitiques Autour de 13 Heures" (Stone-Solid Panics Around 13 O'Clock) and "Croques et Agaceries d'un Gros Bonhome en Bois" (Vignettes and Irritations of a Fat Wooden Blockhead)? And wasn't it Satie again who was found in his room at his death absolutely surrounded by umbrellas, some of them still in the haberdasher's wrapping?

This was Satie. Much of the man's style found its way into the byways of French art, those of Debussy no less than those of Poulenc. Heaven knows what sort of opera Satie would have made of *Pelléas et Mélisande,* for he planned one before ceding the project to Debussy, with advices; certainly the stage works he involved himself with were far more unthinkable before he wrote them (sometimes after as well) than was Debussy's impressionist opera. Satie's *Socrate* (1925), composed in 1919, pleaded a case against program music, for these passages from the *Dialogues* of Plato are accompanied by reiterated contours of sound, unsusceptible of nuance and only atmospherically apropos—no scenery-grimacing or word-pointing in this one. *Socrate* was, as *Louise* had been, a ritual of its time, an obeisance to be made before the caravan moved on. Contained and intermittent, antilyrical, it waded into the ambiguous harmony and parallel motion of impressionism, revealing the text so cleanly one might have been reading.

Simplicity there was, then, in Paris—verbal clarity. Arthur Honegger, reared in France though Swiss by ethnicity (to whatever extent the Swiss may be said to claim such a thing), tackled a number of Biblical and Greek subjects in a style wrought of the impacted largeness of oratorio with the sudden movements of the theatre. In *Le Roi David* (*King David,* 1921), in *Judith* (1926), in *Antigone* (1927), all outward maneuvering is abandoned for directness and thrust. Jean Cocteau's libretto for *Antigone,* derived from Sophocles, is not really an opera libretto at all, but a sessile thing, incapable of movement, and Honegger simply set it as a continuous line of arioso, less melodic than that of *Pelléas* and less fully accompanied. But then *Pelléas* is at heart all romance, while Honegger's stage works are stark and detached, building up to one or two climactic explosions, as when the choral psalms and modal cool of *Judith* pile weight upon weight in ascending to a final scene in which a cacophony of noises heralds Judith's flight from Holofernes' tent, his severed head in her hand.

Honegger, too, tried out comedy, in *Les Aventures du Roi*

Pausole (*The Adventures of King Pausole,* 1930), for comedy is the lure of truth. This was a golden age of one-act plays, slight in premise, rooted in irony, not least because the transparency of impressionist or neoclassical rendition highlighted the work of the poet. No scenery grimaced when Maurice Ravel made his little "comédie musicale" of Franc-Nohain's text for *L'Heure Espagnole* (the title is a pun, meaning both *A Spanish Hour* and *Spanish Time-Keeping*) in 1907, hot on the heels of *Ariane's* debut and fashioned of a similar musico-textual aesthetic but, for a bouffe of no little chic, with the twist of cynical laissez-faire in the action.

First aired in 1911 at the Opéra-Comique, *L'Heure Espagnole* presents a clockmaker's shop in eighteenth-century Toledo, where Madame has to field her garrulous lover, an amorous banker, and a thick-headed muleteer during her weekly assignation while her husband is out setting the municipal chronometers. She hides the first two in grandfather clocks, which the muleteer indulgently shoulders from room to room while her scheduled dalliance goes for naught. Finally, she commandeers the muleteer and arranges with him to come by every week and . . . ah, tell her the time. Slithery string portamentos and dry rejoinders from the pit keep the farce in trim, and further commentary is provided by a takeoff on the old moral epilogue, sung to a cheeky bolero: each character steps forward to call the public's attention to the tropes of jest that they are—the blind husband, the adulterous wife, the fat banker, the babbling versifier, melded, as the muleteer points out, "with a touch of Spain in the air." And, adds the wife, the message is out of Boccaccio: "Among all lovers, the efficient lover wins the day, and a moment arrives in the fortuities of love when the muleteer takes his turn!"

Ravel and Franc-Nohain's bout with the *gai primitif* in *L'Heure Espagnole* points up the "popular" quality of much of the modern era's music drama, for while the composer's depth of musical construction is not aimed at the layman, his emphasis on text gives the work an immediacy that recalls the simple exhibits of the fairgrounds, and the libretto's contribution mates the old intermezzi with the sex comedy of the boulevardier. But then this is the lure of the profane: it is accessible to all. There was excitement in Paris in the 1750s when Pergolesi's *La Serva Padrona* introduced methods of operatic expression not unlike those of Ravel, and so it was again here. The Word, as sound, was coming back into style.

Snazzy, almost jiving with mockery, French comedy faced the sacred plaint of *Pelléas* and *Saint-Sébastien* with the grinning mask, and sentimentalists—of whom there were a few left even in Paris—had little with which to occupy their theatregoing besides the musical comedies of such as André Messager and Louis Ganne. One could count on them for swing duets à la Fragonard, cajoling little confidences, and rousing marches, not to mention a reprise by the entire cast of the show's Big Tune, whatever its literal context. From Spain, however, one receives the valedictory in the realm of the simple commedia neither urbane nor tender—and Spain does take its place in a chapter on French opera because in their slender operatic output the Spanish composers tended to look beyond the Pyrenees for guidance.

It shouldn't have been so, for Spain's was a thriving dramatic tradition, and one would have expected something of an incursion by Spanish opera into the international repertory, especially in the light of Spain's bustling zarzuela industry of musical comedies. But Gilbert and Sullivan, Offenbach, Lehár, and Cole Porter travel about the globe, while the zarzuela is heard only where Spanish is spoken, and the legitimate music dramas travel seldom—not until the impressionist era did Spain even make the musical map at all. There have been hopes of a Spanish national opera, mainly from Felipe Pedrell, composer, musicologist, folklorist, and teacher of, among others, Manuel de Falla, Isaac Albéniz, and Enrique Granados. Pedrell's ambitious *Els Pireneus* (*The Pyrenees*), a trilogy set to a Catalan text, was eleven years old before it saw the footlights in Barcelona in 1902 . . . in Italian translation. Pedrell had envisioned an epic incorporating the detritus of the Spanish cultural foundation—gypsy, Arabian, and Moorish strains vie with the lai, tenson, and sirventes of the Middle Ages in the pages of *Els Pireneus*—and he composed an opera to the text of the classic *La Celestina* (1904), admired but seldom heard.

Spanish to the core, Pedrell belongs among the composers of the nationalist schools, but his students, some of whom paid their scholastic dues in Paris, applied Pedrell's regional ethics to the novel stage experiments typical of prewar France, and for all their locality are French in shape. True, de Falla penned a loud verismatic piece, *La Vida Breve* (*Life Is Short*, 1913), but many of his piano albums tempered the Debussyan prelude with the rhythms of the cubana and the andaluza, and his opera-ballet *El Amor Brujo* (*The Demon Lover*) ratified the "impure" media of the Satie crowd. Granados even adapted two sets of piano preludes into an opera, *Goyescas* (*Pictures from Goya*, 1916), of melodic appeal but orchestrated with the ecstatic militancy of zarzuela.

It was de Falla who came up with the one major opera of this eddy of cultural flow, and that in a bijou barely twenty-five minutes long, *El Retablo de Maese Pedro* (*Master Peter's Puppet Show*, 1923), an enchanting dip into Chapter 26 of the Second

Part of *Don Quixote,* so closely adapted that Cervantes himself rather than de Falla might have prepared the text. This is a complex age, now, not friendly to slogans or round-'em-up labels, but we can press a point and speak of neoclassicism. We've seen it already at work in the rearrival of the Word, an icon of classicism—in Satie's *Socrate,* fussing in its étui for a copy of Plato—but here in de Falla it is less self-conscious, given to the same austere cadences but without the force-feeding of quiescence, and prinked up with a harpsichord in the pit. *El Retablo* is a lively piece, a puppet play about a puppet play, for not only Master Peter's show but the onlookers as well are to be portrayed by marionettes. The spectators include Quixote and Sancho Panza, and, as in Cervantes, the tale enacted is that of the fair Melisendra and how she was rescued from Moorish imprisonment by her husband Don Gayferos.

Much of the score is composed of puppet play mime accompanied by musical interludes, narrated by Master Peter's apprentice, who annoys Quixote with occasional editorializing. When Melisendra leaps onto Gayferos' horse and the Moors pursue them, such is the power of theatre that Quixote intervenes, destroying the entire cast and the little theatre, and he then addresses his fellow spectators with a paean to knight-errantry. Master Peter's comment? "¡Santa Maria!"

A romantic or bourgeois sentimentalist might have tackled "the whole" of Cervantes' epic, as numerous composers did *Faust* in the nineteenth century—and of course we have already had Massenet's *Don Quichotte*—but de Falla, a proper neoclassicist, was exact, and honed in just enough to give full measure. Concision is here, but completeness, too—every note necessary is sounded, no adaptational pass is fumbled, as occurred in that infamous rugby match of a *Faust* opera that Gounod wrote. Less is more is the motto of *El Retablo,* and after some of the more un-gainly place-holding of the post-Wagner era, it is easy to imagine what cool light de Falla's harpsichord must have shed upon the listener. "Boy, stick to your text," Master Peter advises his assistant (in the Ormsby translation). "Keep to your plainsong, and don't attempt harmonies, for they are apt to break down from being over fine."

2. ITALY

It cannot be true, though it is often proposed, that every man has one novel in him, but it does begin to seem as if every Italian composer circa 1900 had one verismo opera in him, because nowadays one each is the sum of their reputations. Even Verdi paid his verismatic dues—when it was nonviolent realism—in *La Traviata,* and Puccini, though often mistakenly lumped with the verists, also took one shot at verismo, in *Il Tabarro.*

As of 1900, with Verdi's career finished, Italian opera was yet at peak activity, Alberto Franchetti, Pietro Mascagni, Ruggiero Leoncavallo, and Umberto Giordano were in their prime, and whereas for continuity France had only the fading Massenet, Germany was suffering from Wagner's failure to be resurrected after all, and Russia lay too isolated to figure much in Western reckoning (*Boris Godunof,* for example, took thirty-four years to cross the frontier, and *Yevgyeni Onyegin* got only as far west as Prague after nine years of wild popularity at home), Italy's operatic balance sheet ran heavily to the black, with promising premieres season in and season out. Unlike French music, however, which was ready to break with romanticism, or at least marry her off morganatically to some new god who would usurp her favors but deny her her rights, Italian sinfonia and opera vocale held on tight to tradition, figuring that renewal would come when it would come.

The practitioners of Verdian romance were most sorely tried, for unlike Wagner, Verdi had no shadows. He had developed too finely to be imitated, and his particular blend of patriotic *sturm* and father-daughter-lover *drang* went limp in less resourceful hands. No, that door was closed; Verdi closed it himself as late as 1893 in a stupendous about-face in mode entitled *Falstaff,* which a Verdi shouldn't possibly have written, and which only Verdi could write.

Ever since verismo had arrived early in the last decade of the nineteenth century, little remained of the grand opera and larger-than-life nobility that Verdi had acceded to in Paris but purified at home. The chorus and its corollary, the massed act finale, were deemphasized for intimacy and precision. No convoy of grandees, pages, monks, or sailors howled when Chenier and Maddalena prepared for glorious death, or when Fedora confronted the man who murdered her lover ("Why did you kill him?" she cries. "For a woman!" he replies. "A woman?" "My . . ." "Your . . . ?" "My wife!" "Tell me everything!" she gasps—*our* sentiments exactly). Moreover, flat recitative had passed from fashion, and while genuine through-composition lay somewhat ahead in Italy, the 1890s broke in several ways of integrating plot business with the Big Tune theory, so as to sing the whole libretto in one language.

One notable holdout was Alberto Franchetti (who almost wrote the opera version of Sardou's *La Tosca,* but was swindled out of the chance when Puccini evinced an interest). Franchetti was the man whom the city of Genoa turned to when seeking a festival piece to commemorate the four-hundredth anniversary of "the discovery of America," and Franchetti and his librettist, Luigi Illica, surfaced with perhaps the last of the big-time grand operas, *Cristoforo Colombo* (1892), a two-part (the Discovery and the Conquest) marathon with enough grandees, pages, monks, and sailors to outfit a Meyerbeer retrospective. Illica was the most industrious of the era's word men, a versatile writer who traversed with ease the grandezza of *Colombo,* the hothouse symbolism of *Iris,* the comic dodges of *Le Maschere,* and the verismo squalor of *Siberia;* he was happy to provide Franchetti with another sortie into historical pageant in *Germania* (1902), a look at the firebrands who attempted to unite the German toparchies into a fatherland. Not so extended a saga as *Cristoforo Colombo, Germania* was set in Germany, but its subject was really Italian patriotism, and in this it is very Verdian. Whether as Sicilians, Flemish, Scotch, or as Jewish captives in Babylonia, oppressed peoples in Verdi are all sons of Aeneas, and though here they're named Loewe, Worms, and Armuth, and though they at one point break into Carl Maria von Weber's "Wilde Jagd" (with Weber himself right on stage, a "pallido giovane" according to Illica's stage directions), the patriots of *Germania* hymn Italy's greatness no less than does the bass-baritone hero of *Cristoforo Colombo.* Pageants of heroism in the twilight of a heroic era, Franchetti's operas were arrested in takeoff by their fulsome conventions—by the expectedness of them.

On the other hand, Italian opera had no intention of retiring the livelier-than-large figures of the nineteenth-century stage and therefore came quite late into the age of satire, discovering a brave new world of symphonic romance and symbolism to explore in the 1910s, 1920s, and 1930s. Then would come breadth and weight, the depth of poetic narration, but at the turn of the century, fast, depictive footwork and naturalism had not yet ceased to be the rage. Credit Pietro Mascagni with the firing of the first bullet in his *Cavalleria Rusticana* in 1890, his stage debut but actually his third opera. Mascagni is one of those composers cited today only, or mainly, for one veristic piece; his crop of sixteen operas, however, yields peacock variety, a survey of early twentieth-century Italian music drama in microcosm, taking in melodrama, chivalric romance, satire, an adaptation of a modish best seller, the inevitable D'Annunzio exhibit, and even profane operetta.

The problem with Mascagni is that after *Cavalleria, L'Amico Fritz,* and *Iris* in the 1890s, the man simply ran out of tunes. Anxious about the machinery of linguistic inflection, he raised the level of Italian recitative, going as far as to write the entirety of the endless *Parisina* in "sung speech," but never was he able to combine his ambitions with the melodic benison that lures the public into moving along with the creator into the next era. So commanding was Mascagni's estate that his and Illica's *Le Maschere (The Masks)* in 1901 received six simultaneous premieres in Rome, Milan, Venice, Verona, Genoa, and Turin (a seventh in Naples came off two days late), but so drained was his muse that the show bombed everywhere but in Rome, where Mascagni's presence on the podium kept things from getting out of hand. No such pressure constrained the audience in Milan, even with Toscanini conducting and Caruso in the cast, and in Genoa the performance was so disrupted that the singers never reached the finale.

This is unfortunate because some of the librettos that Mascagni worked with were among the most potent, artistically, of their day. Illica's premise for *Le Maschere* held that the spirit of the commedia was yet alive, the spirit "that so humanly," as a full company of them sings at the close, "alternates laughter and

tears," so he drew his dramatis personae exclusively from tradition—Colombina, Arlecchino, stuttery Tartaglia, pushy Brighella,
and the Miles Gloriosus, Capitano Spavento (Captain Horrible),
each introducing himself via tiny set pieces in a prologue that
opens with the theatre manager interrupting the overture with
impromptu patter. This personal greeting of the public lacks the
urbanity of the finale of Ravel's *L'Heure Espagnole* (or that of
The Rake's Progress), but then Mascagni hadn't the intention of
undercutting the comedy. His call was to inspire it, to resurrect
the sensitivity of humor as well as the insouciance. Had his melodic gift not gone off a-maying, *Le Maschere* might well have
entered the standard repertory in this modern age of comedy.

—the age as well of medieval romance, these early 1900s: so
along came *Isabeau* (1911), Mascagni's "dramatic legend," set,
according to the poet Illica, in that epoch when "the blossom of
fantasy swarmed over every land, when the hero and heroine
were in flower." This will be "the poetry of the people and the
poetry of the court," he promises, for a love story involving an
overly pure princess and a falconer who is murdered by her subjects when he plays Peeping Tom to her Godiva-like ride through
the city. Determined to push on to a new aesthetic in which music drama could be relayed without the artificial "now song, now
sung-speech" invertebracy hallowed for over two hundred fifty
years, Mascagni arrived at a pliant dramatic arioso for his *Parisina*
(1913), a lurid historical business featuring the most vicious
mother-in-law the stage had yet disclosed in the character of
Stella Dell'Assassino—a contralto, not surprisingly. *Parisina*'s libretto was the work of Gabriele D'Annunzio, a Hofmannsthal of
Italian letters, flavored with Pan and Prometheus. It was he, perhaps, who set the tone for postverismo Italian opera more than
any single composer, for his carnal symbolism and wild-eyed, mi-

natory imagery fired the musicians of his day into deriving a re-
generated poetic theatre just when most other Europeans were
either turning decadent, feeling confused, or simply alienating
the ticketbuyers. D'Annunzio raised Ned with spectators' expectations as to what constituted the drama quotient of lyric drama,
yet he caught their fancy—no!, he fascinated them, demented
them, drove them into myth with his archetypical doom-seekers,
his warriors and viragos possessed by demons shrieking for high
C's and death. Going to the opera on D'Annunzio night was like
enlisting in a dream war, and though one wasn't sure whether
one had won or lost, indeed one felt exhilarated.

It was a national school, really, for D'Annunzio was sometimes an Italian and sometimes a poet but most often an Italian
poet, obsessed with his language, and thus he forced those who
set his verses to follow the shape of the line, to make, as it were,
the sound whole—to redouble the completeness of the script in
the score. Neither in his librettos nor in his plays—trimmed
by the music publisher Tito Ricordi for setting by one of the Ricordi house composers—did D'Annunzio write the sort of beginning-middle-end "bit" that fit so neatly on the recently invented
ten-inch seventy-eight r.p.m. disc. D'Annunzio's characters seldom got the true aria, for even their solo turns were too contextual to travel as facilely out of the theatre and into the recording horn as, say, *Tosca*'s "Recondita armonia" or *Pagliacci*'s
Ballatella. No, they were writing operas in arches now, act to
act to act—or, at any rate, they would do so soon.

Note that Mascagni's dalliance with dramatic arioso parallels
similar developments elsewhere in Europe, some of them called
national and some not, but all of them related to the revived integrity of the Word. Janáček in Moravia, Rimsky-Korsakof in
Russia, Debussy, Dukas, and Ravel in France . . . all were en-

gaged in making their lingual contours an absolute, basing their aesthetic, as the Camerata had based its, thinking it Greek, on nature. From the French came a moto perpetuo of motto themes, from Janáček an angular declamation over germinal motives constantly transformed, from the postverismo Italians rich vocal lines and a rich polyphonic orchestration. But now everyone, not just the Slavs, was brewing a national opera.

Interestingly, the other two Italian composers associated, along with Mascagni, with verismo, were as little inclined to stick to virulent realism as he was. Ruggiero Leoncavallo and Umberto Giordano had both climbed aboard the *Cavalleria* bandwagon in 1892 with *Pagliacci* (*Clowns*) and *Mala Vita* (*Evil Life*), respectively, but Leoncavallo inclined to historical epic and Giordano to costume melodrama rather than the claustrophobic "realism" of the day. Still, they paid their dues to fashion, Leoncavallo with *Zazà* (1900), a study of an actress, somebody's "other woman," Giordano with *Siberia* (1903), which took its soprano and tenor lovers and scoundrelly baritone spy from the gay life in St. Petersburg to the horror of the labor camp—melodramas, theatre pieces run up for the occasion (*Zazà* most particularly because the celebrated Réjane had made a sensation of Pierre Berton and Charles Simon's play). Such entries have short lives, and this *Siberia* was virtually dead on arrival, what with a dismally outmoded evocation of Wagnerian foreboding right out of the *Ring* to open the second act, located at the Siberian frontier. Despite a quotation of "The Volga Boatman," one awaits Fafner.

Both Leoncavallo and Giordano intrigued the ear far more when *not* attempting verismo; almost anyone does. For his part, Leoncavallo, who wrote his own librettos, hoped to found some sort of patriotic panorama, first with a trilogy set in Renaissance

Florence, *Crepusculum* (*Twilight*), left unfinished in the 1890s, and then in a salute to the house of Hohenzollern, *Der Roland von Berlin* (*The Roland of Berlin,* 1904), rowdy with a Meyerbeerian report and briefly in vogue for its plenitude of star arias and a sumptuous gran duo in Act II. Giordano had slightly better luck, latching onto a zesty piece when he tackled Sem Benelli's play *La Cena delle Beffe* (*The Dinner of Jests,* 1924), a sordid binge of trickery in Lorenzo's Florence. The plot hinge is one that any comic opera might have started out with, but Benelli takes it to its logical extreme, extracting the terror from the comical and the grotesque from the familiar. It was a piece ripe for Berg or Busoni; Giordano concentrated on the melodrama, but he did not neglect to abstract Benelli's distorted overturn in his score.

As *The Jest, La Cena delle Beffe* was a smash on Broadway in 1919 with John and Lionel Barrymore as the milquetoast and the bully engaged in role-exchanging and a lethal battle of wits; brother Lionel's nightmarish shriek in Act II, when, chained to a wall, he sees his mistress fondled by John, was the talk of the season. The whole play is pitched at that level, so it should have made a wonderfully bloodcurdling opera, and, in fact, did, being Giordano's best work by far, but for some reason it has never arrived. Subtle as the John Barrymore character is, and brutal as the Lionel Barrymore character is, facing each other off for four acts of ruse and fiendishness, they were born to be sung by tenor and baritone, and Giordano acquitted himself with fervor, making an epiphany of the baritone's exit in Act I—"Here comes death!" he roars, "here comes slaughter!"—and contriving a wan beauty of a signature tune for the tenor, twisted at one spot by a leering dissonance.

Indeed, even as verismo's children applied themselves to

musico-dramatic technique, a transcendently musical era erupted
in Italy, one that finally discovered the Wagnerian orchestra and
the permutations of the leitmotif. Giordano learned little tricks
to do with outlandish harmony, yes, but postverismo opera exploited the magic carpet of symphonic dramatization: Mascagni
wielded stormy psychological intermezzi in *Amica* and *Nerone,*
Puccini similarly defined the morbidezza of *Manon Lescaut* and
Madama Butterfly and painted a nature picture of Rome at
dawn to open the third act of *Tosca,* and Ermanno Wolf-Ferrari
extenuated the ferocity of *I Gioielli della Madonna* with a sweet
little serenade as a prelude to the final act. But the outstanding
such entry in the era was the celebrated Cavalcata in Riccardo
Zandonai's *Giulietta e Romeo,* depicting Romeo's frantic ride
from Mantua to Verona just before the tomb scene. Hearing that
Juliet is dead, he calls for his horse, screaming "Giulietta mia!
Giulietta mia!" The snare drum launches a breakneck canter, the
orchestra leaps into the fray, and the curtain falls, letting the
sinfonia take over, assisted by the chorus, unseen, repeating Romeo's words plus the one other that must be boiling in his brain—
"morta" (dead). How better could the stage play a precipitous
horse ride than by purely musical means? Years later, Benjamin
Britten would cover just such a business (in *The Rape of Lucretia*) in words as well as music on an intimate scale, in a tour
de force of rhythmic and scoring devices and metaphorical commentary, but Zandonai was reaching for scope, for total communion, not for devices and commentary. Lo, the times, the cus-
toms! In post-World War II England, opera must consider its
options, be cunning, have perspective; in post-World War I
Italy, opera charges the battlements of reason, and one is expected to soar along without questioning. That is what the sacred
means.

On the way to symphonic Italian opera, we encounter the man who stood halfway between the theatrical claptrap of the 1890s and the ultramusical epics of the 1910s and 1920s, the unstoppable Giacomo Puccini. Productive if not prolific, ever holding out for a solid libretto, Puccini hit something of a stride after two early flops in the 1880s, and was looked upon as Verdi's heir, though in fact there was little to connect them except success. Moreover, Puccini was the antithesis of the nationalist composer so common among his colleagues. No D'Annunzio symbolism for him, no Risorgimento quest. Puccini's idea of a theatre piece leaned to adaptations decanted from Sardou and Belasco, and as for the fatherland, he preferred to set up his canvas wherever the light led him, be it Japan, the wild West, or old Peking.

Japan, of course, hosted *Madama Butterfly* (1904), with *Tosca* (1900) the most enduring of Puccini's twentieth-century outings, and two more forthright nuncios of romantic melodrama there never were. *Tosca* was based on Sardou's play involving the fiery opera singer, the revolutionary, and the lascivious police chief, dealing all three of them violent deaths, and *Butterfly* hails from the Broadway Theatre of David Belasco, where D.B.'s instinct for kinetic flourish inspired a lengthy "nighttime in Nagaskai" tableau, a series of sound and lighting effects portraying the hours between Butterfly's sighting of her American husband's ship and his presumed return to her the next morning. Puccini retained this in his opera, vivifying it not with stage tricks but with music, and keeping the curtain up at a time when an Italian audience would just as soon spend an intermission at the café in the piazza. *Butterfly*'s hostile opening night public at La Scala took the opportunity to hiss the piece at this point, sending up a cloud of animal noises to match Puccini's bird calls, and not till a few cuts had been made, a few lyrics redrafted, and a second interval added, did the work succeed, but since then its fast ascendance must give one pause. How many operas written between 1900 and the present day would no opera company with a repertory of any size want to be without? One can count the standards in one breath: *Madama Butterfly* and *Tosca*, possibly *Turandot*; *Der Rosenkavalier*, *Salome* and *Elektra*; and what?—*Wozzeck? Peter Grimes?*

Something, in short, must be said for Puccini's staying power, however much he has been maligned by other musicians, and even at worst we are not treating a high-rolling primitive. For all his engagement with popular art—and had not Verdi himself ruled that opera must answer to the eyes and ears of the mob?— Puccini put himself to some effort in *Madama Butterfly* to define Giacosa and Illica's libretto with a consistent tone of voice geared to Butterfly's neurotic belief in her American naval lieutenant and his way of life. Once Butterfly has made her entrance, the music takes on a paralogic intensity, delivered in batches of altered harmonies, "diseased chords," so to speak. The love theme, for example, on its first appearance just a few moments before the heroine steps into view, moves from A Flat Major to an augmented F Major, and continues so in a holding pattern of four whole-tone steps, giving Butterfly's crush on her Pinkerton a monomaniacal character. This is as it should be. Decades before, Wagner had taught the art of giving a distinctive tonal ambience to an opera, in the diatonic choralism of *Die Meistersinger,* in the savage green world of the *Ring,* in the loonybin voyage of *Tristan und Isolde,* with half steps running amuck to uncover the modern concept of antiharmony. Neither Leoncavallo nor Giordano ever sounded like anything but themselves; each opera began where the last had ended. But Mascagni revealed the aural palette to his countrymen, and Puccini was listening; he had heard

Wagner, too, paying homage to *Parsifal's* transformation music in the firing squad march in *Tosca*.

Also like Mascagni—also like most of his contemporaries, be it said—Puccini set himself to reestablishing the position and personality of recitative, the most endemic element in Italian opera, and as far as one can tell the anlage of primeval music drama, chicken and egg at once. In this, however, Puccini never went far enough. Even in his final works there are the full stops, the end of the aria, the pause for public response. Too, there is the theatrical display, the fireworks associated with verismo but plotted into the trajectory of *opera seria* as surely as into the libretto of a *Cavalleria Rusticana*. *Tosca* is particularly clumsy in this, though of course *Tosca* represents costume melodrama, nothing like verismo. It wasn't set in the then-current era, and its realism was the sort that Balzac and Zola would not have recognized; *Louise* is *Tosca's* sister only by happenstance of birth, two weeks apart, and the seamstress' parents would have gaped across the soup at the diva and the Bonapartist as if at beings from Mars. So, perhaps, should we, except that we're so used to Tosca and Cavaradossi now, and we can smile when monographers defame their gambol as meretricious and hypertensive, a "shabby little shocker." *Tosca* is a thriller with a tasty plot, some wildly miscellaneous motto themes that crop up on what can only be called the composer's whim, grains of purposeless farce (the equivalent in *Madama Butterfly,* the Uncle Yakusidé episode, was excised in the second version), and one aria, "E lucevan le stelle," in which the vocal line commences on repeated notes, highlighting the text, while the orchestra dispatches the tune. (So far, so good, but for reasons unknown, this tune is encored at the final curtain, when it couldn't be less apropos.)

Only Richard Strauss and Benjamin Britten compete with

Puccini's staying power in twentieth-century opera, and like
them he didn't vary his structural procedure all that much once
he established a congenial format. Others such as Darius Milhaud and Hans Werner Henze will run a gauntlet of forms later
on, and the scope of their music will change from one form to
the next, but Puccini's development carried through in an exclusively musical passage, and if his musicianship matured, his
dramatic finesse did not—but then, for his purposes, it did not
have to. Whatever flaws one may find in the more extroverted
Tosca, Madama Butterfly has its subtleties, and one will locate
them, always, in the sound of the moment rather than the shape
of the act. Even an "aria in passing" such as Butterfly's "Io seguo
il mio destino" repays investigation for its insistent variations on
the interval of a fourth closing to a third (and for its extrapa-
thetic lines in the unrevised La Scala version: "For me you
spent 100 yen," Butterfly tells Pinkerton; "but I will live with
great economy"). When he wanted to, Puccini wrote with great
economy, as here, making the same statement over and over
again with this figure of fourth and third, as much an obsession
as is Butterfly's beknighted hope for American naturalization.

As theatre people put it, *Tosca* and *Butterfly* "work"; they
play, they do, they do very well. Not nearly so orchestral as their
successors would be, despite their symphonic intermezzi, they
verse themselves in drama; and what dramatic action recom-
mends, Puccini yields, always with an urgency more musical
than verbal. He demanded good librettos, but he carefully
avoided the poetic fireworks of D'Annunzio and kept the enthralled Illica, when they collaborated, firmly on the ground:
the song in the story attracted Puccini more than the story. On
paper, one might raise an eyebrow at the length of the duet between the soprano and the tenor at the end of Butterfly's first act,

but Puccini manipulates the melodic sequences so carefully that the scene flows by as abundance but not hyperbole—an abundance, however, of musical texture. Music, not words, is the information. (This is even more remarkable in *Tosca*, which depends far more than *Butterfly* on the motion of plot: amidst the contortions of *Tosca's* second act, the most perfect moment of remorseless melodrama in all opera, music loses both grace and form keeping up with Scarpia's pincer movements and Tosca's liability, and so must compensate with "Vissi d'arte," also perfect as a specimen of the uncalled-for aria that format jettisoned on the way to Britten.)

So, Puccini . . . a crowd-pleaser with a crowd-pleasing method? Ah, success is always suspicious. There are those who regard it the way a child does a toy: with a desire to isolate the mechanism that makes it work, and break it. The aria? The manner in which orchestral commentary seconds but never overrides the vocal climax? The placement of little solos in passing? What pleases the crowd, which? Anyway, even Puccini wasn't certified as an official crowd-pleaser until late in life, and both *Tosca* and *Butterfly* had less than successful premieres. *Butterfly*, in fact, joins *La Muette de Portici, Tannhäuser,* and *Jonny Spielt Auf* in the precedent of Turbulent Nights at the Opera, for its aforementioned La Scala debut rejected, with the aforementioned crudeness, a work that, virtually unchanged except for small cuts totaling at most three minutes and one new intermission, has since failed to thrill very few.

Along with the cuts in the episodic morsels of ensemble in the wedding scene, Puccini's revision for the Brescia *Butterfly* made an illuminatingly characteristic revision of Butterfly's second-act aria, "Che tua madre"—new words, music little altered. Another of those solos that seems like one of many first-rate tunes

until, like the child and the toy, one breaks it open to find out
why, "Che tua madre" originally referred to a parade of warriors
stopping on the street with the emperor (whom Butterfly addresses, with inadvertent prescience, as "Sommo Duce"), who
admires Butterfly's son and makes him a prince of the realm.
These lines were tossed out for new ones made for the fit, to
center on Butterfly's shame rather than her pride, thus preparing the suicide of the last act. Always, the music counts for
more. Reprising a two-note theme heard at the end of "Io seguo
il mio destino," a 9-8 suspension, Puccini builds "Che tua madre"
on the same figure: pitched in a flat minor, the aria winds around
the same suspension, B♭ to A♭, possibly because it sounded Japanese to him but mainly because these two notes are identified
with Butterfly and crop up repeatedly in her music, from the last
two notes of her entrance (her "Giù" is the 9, her friends' "Giù"
the 8) to the final bars of the opera, when the orchestra thunders
out the melody of "Che tua madre" one last time. Puccini is not
comparable to Strauss or Britten in musical technology, but,
clearly, he is a man whose resources are easy to overlook.

It cost the Metropolitan Opera $22,800 to secure the rights to
the world premiere in 1910 of Puccini's next work, *La Fanciulla
del West* (*The Girl of the West*), not to mention the fees of
Emmy Destinn, Enrico Caruso, Antonio Scotti, and Arturo Toscanini; by the time the curtain fell it was generally conceded
to have been a bad investment. Forever after, this *Girl of the
(Golden) West,* another borrowing from Belasco, has invoked
groanings and gigglings rather than a raising of the hair, but
while it is a less than authentic dip into American folklore, it
ranks high in the composer's canon for its adroit use of impressionism, not only in the tergiversating harmony but also in the
rapport between vocal and orchestral melody.

The fanciulla, yclept Minnie, is the proprietress of the Polka saloon in the days of the gold fever, schoolmarm, mother surrogate, and tomboy to the rough miners who frequent the Polka. Nasty sheriff Rance (baritone) loves her; so does Dick Johnson (tenor), alias Ramerrez, an outlaw leader. Things come to a head in a poker game, "due mani sopra tre" (two out of three), to decide whether Minnie or Rance gains custody of the wounded Johnson. If Minnie wins, Rance must clear out, but if Rance wins, he gets not only his man, but the woman as well: that sort of opera. Lucky in love, unlucky at cards—and Minnie is about to lose the deciding deal when she slips an extra ace into her hand. A splendid scene, this, replete with parlando, and if its mingling of western shoot-'em-up tropes and Italian melodrama comes off as inadvertently droll, what western, what melodrama, doesn't?

Puccini meant these whole-tone doings out on the range as a legitimate entry in the widening horizon of his native art form, but the entrance of the miners into the saloon with their Hello's, with the bartender's "Buona sera, ragazzi," and with two men singing "Dooda, dooda dooda, day" gets the evening off to a dubious start, even after the thrust of the tiny prelude, a study in impressionist chording but filled out with the energy of Latinate canto. Puccini seldom resisted the chance to regionalize his scores with folk effluvia—*Madama Butterfly* is studded with pentatonic themes, genuinely Japanese ones, albeit harmonized for Western sensibility—and *Fanciulla* really does convey some aroma of the nugget-hustling Sierras when the orchestra has its say. An ingenuous clopping figure imitative of a galloping horse for the arrival of the Pony Express is a mere blandishment, but the tenor's "Quello che tacete," used in all three acts, is branded with the unmistakable ring of American folk song. Best of all, *Fanciulla* shows Puccini entering his prime as a word man, communicating through his setting of Guelfo Civininni and Carlo Zangarini's lines, but in the silences between them as well. The finale of Act I, a dialogue between Minnie and Johnson, is a trio for soprano, tenor, and orchestra, eloquent of both fumbling shyness and yet of Minnie's expansive spirit. The curtain drops on one of Puccini's masterstrokes: "You don't know yourself, Minnie," says Johnson, while fifteen tenors, offstage, hum the theme, built of major sevenths, associated with her. "You're a creature good and pure in heart," he tells her, "and you have the face of an angel." He leaves Minnie in the middle of the empty, darkened saloon, trying to make sense of what she has just heard. Slowly she moves downstage until, almost at the curtain line, she repeats Johnson's last words, "the face of an angel," and covers her face with her hands, the schoolmarm, mother, and tomboy smarting from the pain of turning into a woman—and as the curtains slowly close on a thunderous echo of her theme, she sighs.

Too concerned with standard romantic maneuverings ever to assault the heights of the sacred, Puccini blundered in the opposite direction with his would-be operetta, *La Rondine* (*The Swallow*, 1917), designed as a musical comedy in Viennese style but ultimately unveiled as nothing more than a thin little opera swooning with waltzes. Mascagni delved into operetta around this time, in *Sì* (1919); he came closer to the mark, treating a libretto modeled on *Der Graf von Luxemburg*, whereas Puccini's piece hewed more to *La Traviata* than to Léhar. *La Rondine*'s heroine, Magda, lacks the earthy glow of the light comic stage, and her chambermaid is portrayed in unconvincingly mercurial cacophony, but Mascagni's protagonist Sì (so called because she never says no) managed to convey the air of the music hall with aplomb.

Puccini stepped into the profane with gusto in the third part of his bill of one-act operas, *Il Trittico* (*The Triptych*, 1918). This was *Gianni Schicchi*, a riotous caper with the sardonic reprobacy of *Falstaff*, derived from a hint in the thirtieth canto of the *Inferno*. Giovacchino Forzano's libretto about the upward mobility of Florentine know-how bore a resemblance to the old commedia, nimbly dramatized by the composer in what was for him a stylistic breakthrough, what with hypocrisy and cowardice dripping buoyantly out of the score as the upstart Schicchi colludes with the relatives of the late Buoso Donati to impersonate Buoso and dictate a new will. One gaffe intrudes on the merriment, the aria "O mio babbino caro," a throwback to morbidezza, as if in the protective coloration of a threatened species of butterfly—or was Puccini perhaps spoofing himself? Certainly there is no mistaking the self-mockery of the opera's finale, when Schicchi, having impersonated a dying Buoso to leave the bulk of Buoso's fortune to a certain Gianni Schicchi, turns to the audience and, speaking, admits that though the great Dante has consigned him to hell for his crime, he hopes for the theatre's time-honored reprieve . . . a little applause, ladies and gentlemen?

Il Trittico is a rather mixed bag; no overriding concept associates it, and its parts are seen more frequently on a bill with some other work than, as intended, in trio, most likely because Part II, *Suor Angelica*, a treacly tale set in a nunnery, does not equal its brothers. Part I, *Il Tabarro* (*The Overcoat*), was adapted from a Grand Guignol favorite, Didier Gold's *La Houppelande*, and constitutes Puccini's sole verismo excursion. Laid in the claustrophobic slums of Paris, on a barge in the Seine, *Il Tabarro* reprised the impressionistic coloring of *La Fanciulla del West*, especially in the opening water music and, with *Schicchi*, is occasionally nominated as Puccini's masterpiece for its wiry economy.

But no. For the masterpiece, one must consider *Turandot* (1926), looking past the "modern" orchestration and the unexpectedly Wagnerian demands of the title role to ponder point-of-view. This is a problematic work, not least because Puccini left it unfinished at his death in 1924. Franco Alfano, who had proved his facility for exotic orchestration in *La Leggenda di Sakùntala* in 1921, was entrusted with the sketches for the last half of Act III, and the popular canard has it that Alfano illy mixed his grout and marred the mosaic. On the contrary, Puccini left extensive notes and Alfano did well by them—better than appears today, for the first edition of the score tendered a more detailed working out of Turandot's transformation and a slightly less bald restatement of "Nessun dorma" in the final chorus. The problem is not that the psychological emergence of the impossible *Turandot* was botched, but that a psychological development was attempted at all, for Puccini's *Turandot*, like Busoni's, is a fairytale, not a Freudian study, and like most fairytales it is unreasonable, arbitrary, and relentlessly nihilistic.

The libretto by Giuseppe Adami and Renato Simoni does not follow Gozzi as closely as Busoni did, but the source is Gozzi, and Gozzi's the tone. We are again in legendary Peking, again in a world where princes fall besottedly in love at first sight, where princesses require a courtship of three riddles of their gentlemen callers, beheading the slow-witted, and where the suicide of a lovelorn slave prompts a choral lament but little or no empathy from the principals. Drawing on the pungency that kept *Gianni Schicchi* afloat, the authors retained three of Gozzi's commedia courtiers, dropping the traditional Italian names for Chinese monosyllables—Ping, Pang, and Pong—and Puccini rendered much of Gozzi in musical language, with a chorus that veers from bloodlust to pity in a trice, laying misshapen motives side by side with Big Tunes, and using the three comedians to spoof the whole business . . . sometimes. As it happens, the composer could not resist giving their scene-long trio in Act II moments of inapposite sincerity, but "let's go enjoy the zillionth torture!" they giggle as two unmuted trumpets call the court to the riddle contest.

Still, if Puccini's *Turandot* lacks the singleminded execution of Busoni's version, it remains a unique work even for the Italian 1920s, when fable and fantasy predominated almost to the exclusion of anything else. Many finely wrought operas of its time have gone under, but *Turandot* survives—no small testimonial to its at times ugly power. In Busoni, as in Gozzi, Turandot's resolve crumbles as soon as she claps eyes on the Unknown Prince, but in Adami and Simoni's scenario she is a figure of "ice and fire" right up to the final confrontation with her conqueror and his kiss. When the public first encounters her, she doesn't sing a note, a defeat of cliché if no real novelty, and when she next appears it is to launch immediately into the Annapurna of soprano arias, "In questa reggia." Her air of remoteness, of quasidemonic possession, is doublestopped in the riddle scene, more rhythmic dialogue than song and built on one three-note theme (and bequeathing another set of riddles, the answers being hope, blood, and—"what is the ice that gives you fire and from your fire derives more ice?"—Turandot herself), and if Puccini mitigated the inscrutability of his fable with the all too scrutable slave girl, Liù, his construction of this strident, volatile faery neverland was his greatest achievement.

Polymorphic opera, hydra-headed, cannot be expected to progress from here to there like an army, and it is instructive to see

how different Puccini's *Turandot* is from Busoni's, only eleven
years its senior but already receptive of the neoclassical witness.
In Puccini, who could take novelty just so far, the scenery grimaces. *Turandot's* oriental setting and nightmare atmosphere led
him to take drastic steps in scoring (the influence of Stravinsky
has often been noted), while for Busoni it was a thin flash, the
light touch, gestures.

Light, as well, was Ermanno Wolf-Ferrari, who shared one
point of reference with Busoni, German-Italian parentage. Born
in Venice, Wolf-Ferrari was in art all Italian. Of his thirteen operas, seven debuted in Germany in German translation, but almost all take place—and root—on home soil, and the gentle shade
of Goldoni, not of extravagant Gozzi, dictates the tone. As if to
prove that he could, as if all of his generation had to, he duti-
fully handed in his verismo assignment, *I Gioielli della Madonna*
(*The Jewels of the Madonna,* 1911), but no one of his arena
played commedia so refulgently as he did; a new Mozart, some
said.

It almost seemed so. Supple, melodious, swift, or patient, depending, Wolf-Ferrari's comedies restored the innocence of Italian farce much more easily than Mascagni was able to in *Le
Maschere,* by simply throwing out the masks (as Goldoni had
done) and letting natural uproar evolve as, nearly, in life. It's
hard to believe that *I Quattro Rusteghi* (*The Four Babbitts,*
1906) received its first hearing in German, so idiomatic is the
composer's presentation of Giuseppe Pizzolato's Venetian dialect, as luxurious a medium as the mud that layers the shallows
of the lagoon. (Nubile daughter to stepmother, about her coming
marriage, for "Tell, tell me, is anything decided?": "La diga, la
me diga, ghe xe gnente in cantier?"—as much Spanish in it as
Italian.)

I Quattro Rusteghi, as a prototype of twentieth-century comic opera, lacks the burnished ecumenism of *Die Meistersinger* and *Falstaff* and the unstinted melodic congruence of Mozart—and never could the Venetian deal in Busoni's polytonal harmonics, pedal points, or inverted recapitulations—but as Wolf-Ferrari's best of several first-rate farces, it captured well the egocentric rodomontade of commedia, squarely in the tradition rather than breaking into seriocomedy. Another "one mad day" in the life of, like *Le Nozze di Figaro*, *I Quattro Rusteghi* starts in the morning when little Lucieta hears that she is to be wed, but that her boorish father and his three boorish friends have decided not to let her meet her intended until the ceremony. The wives enter the picture that afternoon with a bit of intrigue, outraging the husbands, who call everything off, but by that evening all is put to rights and everyone goes off to dinner.

Wolf-Ferrari did not attempt to turn this Goldoni-derived tale into anything very grand; it's a winsome piece with some flavorful vocalism and charm to spare—not least when, at the start of Act III, the curmudgeonly *Rusteghi* sing, talk, and shout diatribes against their mutinous wives. There's a strong feeling of Pergolesi about here, as there was in Wolf-Ferrari's other comedies drawn from Goldoni, *Le Donne Curiose* (*The Curious Women*, 1903), *La Vedova Scaltra* (*The Wily Widow*, 1931), and *Il Campiello* (*The Square*, 1936). Moreover, *Il Segreto di Susanna* (*Susanna's Secret*, 1909), though it proffered the modish dainty of a woman possessed by the temptation of the cigarette (her husband thinks she's deceiving him with a lover; their reconciliation is irresistible), harked back to the ancient intermezzo in its three-character, single-action brevity.

But era has its advantages, and in at least one work, *Sly* (1927), subtitled, "the legend of the sleeper awakened," Wolf-Ferrari moved into the seriocomic world, rehabilitating an old touchstone situation with the tactics of transition. The supposal:

a drunken buffoon is carried, asleep, to the house of a lord, to be convinced that he is a wealthy magnifico and then ridiculed when he accepts the ruse. The librettist was Giovacchino Forzano, who had learned much from the masquerade gambit of *Gianni Schicchi,* a lighthearted study for this cruel *Sly* (originally intended for Puccini, in fact). Forzano borrowed his antihero from the Induction to Shakespeare's *The Taming of the Shrew,* but since Shakespeare's resolution of the situation is no longer extant, Forzano was free to bring the hoodwinked Sly to his knees like a Wozzeck, not merely fooling him with the joke, but convincing him that he is the lord of the manor so irrevocably that rather than give up the dream when it is shattered, he cuts open his veins with a shard of glass.

For this bimodal proposition Wolf-Ferrari had to secure a comic voice less convivial than that of his Goldoni adaptations. *Sly* called for the gay and the ghastly at once, and in this the composer was a somewhat less than total conquistador, for much of *Sly* leans to the glib horror-music of verismo. Still, in Sly's own music particularly, Wolf-Ferrari vised the fingers of bourgeois comedy and waved them in an air as much shadow as light, responding to the rage for problems of human identity rather than domestic intrigue in the theatre. The protagonist's "I am not a buffoon!" appears especially despairing after his earlier strophic narrative, the bumptious Song of the Bear, and let us sample a dram of Sly's apostrophe to wine in the tavern of Act I, when his life is still a simple matter of being drunk or sober ("Drunk am I?" he cries. "No—intoxicated!"):

> *Ma bevi, bevi!*
> *Quando non bevi,*
> *Chi sei povero Sly?*
> *Cantore di taverna . . .*

Giocoliere . . .
Venditore ambulante . . .
Scrivi versi . . .
Ma puoi fare di tutto,
Tu resti un miserabile!
Uno straccione!

So drink, drink up!
What are you, poor Sly,
When you don't drink?
A tavern singer,
A juggler,
A peddler,
You write verses . . .
But no matter what you do
You remain a wretch of a man,
A scarecrow!

Wrenched from antic to romance, the scarecrow brought to life and treated like a nobleman, Sly cannot release himself from the dream, and when the idyll ends, Sly goes with it.

Said Oscar Wilde, "History is gossip"—what an absurd reduction. History is a lavish catalogue of double toil and trouble, most lavish when it would, like Shakespeare's Hecate, "show the glory of our art." This chapter, of all chapters in opera, is the lavish one, for there is more to tell here in Italy than anywhere else, from verismo and Sardouvian melodrama into symbolism and legend, with satire snapping at D'Annunzio's heels, hungry for dinner. In Richard Strauss we will encounter the most successful of the prolific composers and in Alban Berg the most acute of the moderns, and both Benjamin Britten and Hans Werner Henze await us at the far court, but it is the lyric theatre of Verdi's heirs that most ignites the imagination, that in greatest quantity reheats the opera stewpot with ambitious minestra. Except for Gian Francesco Malipiero, neoclassicism belonged to the north through the 1930s, and the savage transformations of tragic satire were worked in Teutonic circles, but Italy held out for a golden age of romantic opera, more musical than literal and posed for tableaus of the story-ceremonial—the doomed lovers, the heroic initiation, the separation and return.

Luckily, Italy had reared an amazing generation of actor-singers by 1910. This is all memory now: the Italian companies regularly import American sopranos to sing the top line in Verdi—and what Italian soprano since Gina Cigna has been able to negotiate the riddle scene in *Turandot?*—but in the age of Franco Alfano, Italo Montemezzi, Ottorino Respighi, Riccardo Zandonai, and Ildebrando Pizzetti, the national opera held out as a national preserve. Such names as these haven't the fillip of even limited familiarity, it's true, and in the aggregate they may sound like a footnote on Justly Neglected Works by Minor Masters . . . but remember, the word "minor" comes easily only to those who create nothing themselves, and as for justly neglected works, that's exactly what *Les Troyens* was thought to be a prime example of twenty-five years ago. Bear with these strangers and their unknown princes of opera: their day will come again.

And what a treasury of talent there was on call and at the service of almost monthly mountings of new works in the 1910s, 1920s, and 1930s. Think of the sopranos alone—Dusolina Giannini, Carmen Melis, Rosetta Pampanini, Gilda Dalla Rizza, Bianca Scacciatti, Mafalda Favero, Lina Bruna-Rasa, Toti Dal Monte, Maria Caniglia. Some of them shrilled on the top notes, others developed chronic wobble, and a few were totally unsuited to parts they insisted on singing—and which they grew into so organically, like Dalla Rizza as Violetta Valéry, that eventually they came to own them—but in all it's a formidable group, complemented by a pride of mezzo-sopranos, tenors, baritones, and basses, the troop of them devoted to, honestly, art. Could it be that this richness of supply created a demand for as many intriguing new works as could be fitted onto stages?—for certainly Italian opera has never since intrigued so thoroughly. There is a paucity of tophole Italian singers nowadays, and likewise a paucity of important new operas.

But what a business it was then, what a source of headlines! Even the oldsters of the verismo generation rallied with entries bent on redeeming their first flash successes back in the 1890s. Leoncavallo tackled Sophocles' Oedipus to his own text, unfortunately bringing him to the ground, as *Edipo Re* (1920), but Mascagni's *Il Piccolo Marat* (*The Little Marat*, 1921) proved that neither he nor his librettist, Forzano, was willing to coast along on convention. A gruesome tale in which, oddly enough, the good end happily and the bad unhappily, *Il Piccolo Marat* was set in the French Revolution, typical operatic decor here utilized quite untypically.

But it fell to the younger generation to deliver the gold, and it was they who greeted the sacred, if seldom the profane, in a rota of more or less symbolist pieces of a striking sincerity, fertile with orchestral matter and forceful, poetic librettos. Inheriting the constructions of late-middle Verdi for *Giovanni Gallurese* (1905), a look at Sardinia under Spanish tyranny in the 1600s, Italo Montemezzi then tossed out the old last for *L'Amore dei Tre Re* (*The Love of Three Kings*, 1913), a setting of Sem Benelli's play, reduced by Benelli himself. The materials, on the surface, could have served Verdi, or even someone like Filippo Marchetti, but it is how these materials are plied that marks the new school, and in certain cases the tracks left by *Tristan* and *Pelléas* are metres deep.

Benelli laid his tale in a "remote castle in Italy during the middle ages, forty years after a Barbaric invasion." The three kings of the title are Archibaldo, the man who led the invaders two-score years earlier, now old and blind; his son, Manfredo, the present king; and Avito, the rightful pretender. Their love is Fiora, the embodiment of Italy itself, beautiful, fiery, and capable of great tenderness, even to her subjugator. In music of an extremely evocative nature, wildly rhythmic and, in the love scenes, boiling with a chromatic impulse scarce heard before on the Italian stage, Montemezzi created one of the masterworks of the period, a theatrical symphony with few of the closed forms and arioso-to-aria gear-shifting of Puccini's operas.

There is only one aria as such in *L'Amore dei Tre Re,* right at the start of the evening, "Italia! Italia . . . è tutto il mio ricordo!," Archibaldo's reminiscence of his rape of "this goddess born between two seas." Oboes and horns describe a martial foray while the strings rush in arpeggios and Archibaldo smoothes over the carnage with a sweeping tune, all of this immediately routed by bursts of dramatic narrative, and this in turn broken into by a coarse motive in the lower strings. One tune, then a second, and a third, hunt one another as barbarians hunt, shooting back and forth helter-skelter in polyphonic strands to a sudden climax. It is a defilade of keys and moods, and, in effect, no aria at all, not the sort easily shared with the gramophone but rather synergic and contextual, as necessary to the drama as the rise of the curtain.

So integral a framework as Montemezzi's has its drawbacks in performance, alas, which is one reason why he and his colleagues of the postverismo reclamation are not often encountered today, for they demand not only strong acting-singers but brilliant conductors, not the staff drudges usually assigned to them. This is, above all, a *musical* era, snubbing the Word-revival of neoclassicism even as it barges into poetic realms, and these narrative tone poems are labyrinths of technique. A dreary conductor can mar but not destroy *La Traviata* or *Rigoletto,* but a weak baton dissolves Montemezzi and Benelli like the touch of lime: the music sags, the poetry drivels. Great conducting of this repertory is pure lightning, however, as Montemezzi himself proved in the pit of the Metropolitan Opera production of *L'Amore dei Tre Re* in the early 1940s with Grace Moore and Ezio Pinza.

Montemezzi's seldom if ever performed *La Nave* (*The Ship,* 1918) continued his investigation of barbaric passion, this time to the verses of Gabriele D'Annunzio, reduced from playscript to libretto by the omnipresent Tito Ricordi. Difficult to cast and of astounding proportion in the way of physical production, *La Nave* was doomed to depart the scene early on—a vexatious loss, for it must be the most colorfully sadomasochistic opera in the annals. Set in the Venetian lagoon in the days when it was barely inhabited, long before the city usurped the seas, *La Nave* is saturated with nautical imagery, including an ocean motive used throughout, as in Rimsky-Korsakof's *Sadko,* a kind of preceptive ostinato. The two "romantic" leads, soprano and tenor of unflinching larynx and hottentot charisma, are characterized as epic avatars, driven by demons of power and lust, sworn enemies, sworn lovers, a pair of Attilas. The passionate sweep of *L'Amore dei Tre Re* works as well here, and D'Annunzio's text, though unremittingly revolting, is in its heroic lyricism a paradigm of the symbolist libretto that must have music. For the Italian nationalists, feats of internecine ruthlessness were as much a part of the telling tales as was chauvinist splendor; D'Annunzio treated his public to a sword fight to the death between two brothers, plus the heroine's brazen execution of prisoners, one by

one, by bow and arrow, along with some lascivious installments
of the two principals' love-hate affair, and the scent of coming
glory as the improvised lagoon colony makes its first piratical
claims on the trade routes of the Mediterranean.

Speaking of barbarians, World War I almost literally wound
up on *La Nave*'s opening night at La Scala, the happy news de-
laying the curtain for an announcement of the capture of Trieste
by Italian troops, which made D'Annunzio's bellicose libretto
very much of moment. The eponymous warship itself is the Totus Mundus (literally "All the World"), for the authors were
hymning not just Venetian sea power but the universal drive for
possession that tempts sopranos and tenors as well as nations.
The work's final tableau counts as one of the most awesome
sights in all opera (or, rather, it would so count if anyone were
willing to revive this gorgeous white elephant): in the moments
just before dawn, the tenor decides to sail off in the Totus Mundus and conquer the "New Rome"—"Settle all the Adriatic for
the Venetians!" cries the mob—but before he leaves, he lashes
his beloved nemesis to the prow. "Faledra," he assures her, "I
give you a beautiful death." Then the props are released, the ty-
ings burst, the galley oars rise, and as the mob howls and the
sun comes up, the ship glides majestically into the foam like the
wrath of God.

The monolithic was the rule in this corner of opera history—
hugeness of spirit rather than simple spectacle—and those who
could not compete on these terms tended to sound like molly-
coddles. Legend arrested this generation, fledged it for headlong
flight, and committed it to the sacred, pushing middle-class opera, the Puccini idea, off to the side. They sought out eternal images, something out of time, not out of Sardou. That immense
warship heaving into the sea in *La Nave* was such an image—
theatrical, yes, but something more, charging the spectator to see

past the stage and singers into the Italian past and future, seducing them into an almost holy communion, with the fatherland as God. On a less epical level, Sem Benelli gave Montemezzi one of the pictures of the age in Act II of *L'Amore dei Tre Re,* when Fiora's barbarian husband sends her a long white scarf which she is to wave at him from the battlements of his castle as he rides off to war. To hot-blooded surgings in the orchestra she accedes to his request, floating the veil in the air again, and again, and again, while a second "husband," the true Italian king, waits nearby hoping to take her away with him, and a third "husband," her blind father-in-law, hovers somewhere in the vicinity, minutes now from strangling her for betraying his son. This is Italy—wife, daughter, and lover, handed from one rude conqueror to another. Scarcely finished waving her chaste white marker, Fiora turns to the Italian heir, whose importuning she can no longer withstand. "Ah!" she cries. "I yield my rights as does the pitying tree to him who dies of thirst." "I thirst," replies her king. "I thirst!" And the orchestra explodes.

Will we ever see the like again? This was the last hour of untrammeled operatic romance, before total politics, total communications media, and total Godotism robbed us of the unashamedly heroic, perhaps even of belief in heroes. The masterpieces are few enough now, but they were regular events then. Just a year after *L'Amore dei Tre Re* came Riccardo Zandonai's *Francesca da Rimini* (1914), its libretto lifted from D'Annunzio's play, written for, and by all accounts about, Eleonora Duse, his light of love. Duse toured it all over the Western world, and just when it showed signs of fading, Zandonai's opera arrived and the tale burned red all over again, this time without Duse but with music, ecstatic and stupendous music, a dream, a nightmare, a dream.

As Vera Stravinsky once said (in regard to the world premiere of *The Rake's Progress,* in Venice), nothing important in Italy begins on time, and here was Italy's most fulfilling answer to Wagnerian love-death, fifty years late. Francesca and her brother-in-law Paolo, of course, are the woebegone adulterors from Canto V of the *Inferno* ("Nessun maggior dolore . . .") who were reading in the Book of Galeotto when temptation overcame them, "and that day we read no more." This misguided romance finds its mirror in the music in that Zandonai lowers the dominant seventh a half-step when it resolves to the tonic, a somber, compulsive sound for risk-takers; here are lovers fated by the largeness of their own conception to immortal but not much mortal life. "How would you have me die?" asks Paolo of Francesca at one point, and she: "Like the slave at the oar in the galley named Despair—so would I have you die!"

Francesca da Rimini is continually mistreated by critics, but scarcely a year goes by without some Italian company mounting a production, and even abroad it has not suffered the complete disuse that has waylaid most of its coevals from the symphonic renaissance of post-Verdian Italy. Less raucous than Zandonai's *Conchita* (1911), a sleazy verismo piece, and less copious than his *I Cavalieri di Ekebù* (*The Knights of Ekebu*, 1925), an unexpected adaptation from the novel, the source being Selma Lagerlöf's *The Story of Gosta Berling, Francesca da Rimini* marks a kind of fulfillment of late romantic romanticism, as passionate as anything in Verdi but scored for vast orchestral strategy. The great meeting of Francesca and Paolo at the end of Act I, a twist on *Der Rosenkavalier's* betrothal ceremony, is a moment of archetypal immensity, all the more vivid because we have already got some measure of the fatalistic heroine but know nothing of the hero other than that his extreme good looks are being used to trick her into marriage with his older brother.

This is the stuff of great poetic device, moving hero and heroine into place so the maddening night of their last days can commence. Paolo meets Francesca in a lie, and the two will live a lie until Francesca's husband catches them together and murders them both. *We* know this already, of course, through Dante if not instinct, but words and music both are so outrageously dramatic that this prototypal adventure plays as a first-time experience. Zandonai builds up to Francesca and Paolo's meeting in a suspenseful series of movements, *allegro brillante e*—significantly—*deciso* ceding to *sostenuto* and back to the *allegro* as serving maids scream for Francesca to come greet her supposed fiancé while Francesca veers between resignation and defiance. Throughout, a diatonic four-note "fate" theme obtrudes insistently, goading the action, until a crescendo explodes it into pianissimo arpeggios of its constituent harmonics, *largo*: Paolo has appeared and Francesca stands transfixed. Now comes the love theme, an ecstatic line in D Major, articulate with yearning but striving for control and pinched at moments by remote tonalities. "Played" by onstage musicians, it is scored for a medieval character, and as the full orchestra rears up with it, Francesca plucks a rose and offers it to Paolo. The music begins to fade on repetitions of the fate theme, but the last quotation suddenly rises up to an altered second, stabbing the dream. Surely they know, both of them, how it must end.

This is not, nor was meant to be, reheated leftovers from the nineteenth century; either Berlioz or Wagner might have treated the tale with the requisite poetic interior and a comprehension of the tragic anima that craves the long night. But neither treated the subject in any way at all, and Ambroise Thomas' stately *Françoise de Rimini* of 1882 only demonstrates how much more fully opera in the twentieth century could justify men's

ways to men, as does Zandonai in his whack at another touchstone subject, *Giulietta e Romeo* (1922), a tightly knit, intimate,
and antierotic tale, stripped of epic incidentals. Having studied
his Shakespeare, not to mention Masuccio, Bandello, and Giro-
lamo de la Corte, Zandonai's librettist Arturo Rossato wrote his
own lines but retained the spirit of the Bard (surely the best
reading *that* tale has ever had), his fever, his shyness, his fear
of touch. Shorn of the senior Capulets, the Nurse, Friar Law-
rence, and Mercutio, *Giulietta e Romeo* focuses on the central
love affair and the insensate feud that destroys it, and never stops
singing. For *Francesca da Rimini,* Zandonai had borrowed the
lute, fife, and viola pomposa of the Middle Ages to evoke, besides time and place, the credence in storytelling and ballad
forms that we no longer have the time to subscribe to; similarly,
in *Giulietta e Romeo,* the fatal turn of plot—Romeo's accidental
hearing of Juliet's apparent death—is delivered by a minstrel
parading his latest ware, the ballad of this same Capulet daughter.

This medieval pastiche is neatly blended into the score, but
be warned: we are approaching a day when self-conscious quotation, unblended, will be welcome—a day when little forms
tucked into larger will tame romance with calculation, a day that
will see past the well-made play, and the well-made libretto,
partly to reinvent the structure of opera. The neoclassical revival
has much to do with this, as it sweeps the terrain with a net for
parodistic forms, and its concomitant, absurdist comedy, will
blast beginning-middle-end narrative off the stage. Some librettists, now, will offer pieces without a beginning, some pieces
without an end, and a very few others will prance about in
naught but a vaporous middle. The spoken drama picks up
quickly on new modes, opera more slowly, but there is much

collusion between the two, and while Italy held on longest to the straight narrative format, it too will yield.

It was already yielding, to an extent, in the work of Ildebrando Pizzetti, who after the indispensable D'Annunzio entry, *Fedra* (1915), took to writing his own words for original works conceived organically for operatic shape and bearing no vestige of intermedium identity. A musician of the sort who impresses his confraternity more than his public, Pizzetti was determined to redeem music drama of the mortgage by which the music held the drama in forfeit. Retaining the solo when necessary—and only when—Pizzetti disconcerted the aria, devising instead a fluid, adaptable vocal language sculpted so coincidentally with the orchestral parts as to make of this ancient word-music controversy an old bone, something for museums.

Mascagni's solution had taken the form of a tedious arioso, while Puccini, when not writing wholesale arias, hit on a compromise that favored either the orchestra, as in *La Fanciulla del West* particularly, or the voice, but in the 1920s the integrity of through-composition ceased to be an issue. Thus came Pizzetti's monolithic *Débora e Jaèle* (1922) and *Fra Gherardo* (1928), an astounding pair suggested by the Bible and medieval chronicle, darkly nuanced and reasoned out, tapestries spun of granite. With no literary source to placate, Pizzetti constructed operas to play through music more than through stage pictures and certifiable Moments, and his chorus-writing is especially remarkable.

Débora e Jaèle, for example, opens with a lengthy, intricate picture of Israelites terrified of oppression by the Canaanites, quarreling and howling for their fatidic leader, Debora. Themes that later implant themselves as leitmotifs well up and dissolve in a polyphonic bazaar, now in the choir, now in the pit, tendrils winding about each other as if to strangle themselves—but on the contrary, the mass breathes. Ranging at times hysterically from the cool, open chords of diatonic purity to banshee chromaticism, the scene builds to the appearance of the prophetess on a thundering C Major, needled, as the crowd falls to the ground, by pianissimo seconds, fourths, and flatted notes: a very legitimate magic.

Fra Gherardo, the crown of Pizzetti's youth, tendered more color than had the Old Testament gravity of *Débora e Jaèle.* Set in seventeenth-century Parma, *Gherardo* traced the rise and fall of a weaver turned fanatic monk who takes on the churchly state and is ruined at last by his rage for truth. The rabble is even busier here than it was in *Débora,* its music laden with a rapturous ferocity that leaps, from life, off of the mean streets and onto the stage. Such complex artwork as Pizzetti's is not the kind that wins international or even local currency. There is too much to take in, too intellectual a communion beseeched of the spectator. Not surprisingly, the most popular scene in all of *Fra Gherardo* was the hero's recitation of the Magdalene story, "Un giorno Gesù stava a mensa," a light strophic solo in folkish triple time—a Tune, as it were.

The sum of all this is the sacred, communicative of shared symbols, seeking musical remedies more often than not, fleeing from realism into nationalism and heraldic largesse. Even in Pizzetti's declamative tragedies, one is less conscious of the Word than of its musical expression, and this is after all an imperiously symphonic era, friendly to characters larger than death. "My enemy was daylight, my friend was night," states D'Annunzio's Paolo of his exile from Francesca in lines not from the play but written especially for Zandonai's opera. The nonmusical theatre did not have to dwell in Ultima Thule, but opera often makes the voyage—music takes it there—and what it finds it raises up to heights. Everything takes on a sheen, and in this particular Italy, the glow was late romantic, mannerist but robust. Even in Franco Alfano's one-act *Madonna Imperia* (1927), based on one of the more licentious of Balzac's *Droll Stories,* and revealed in a dextrous conversational recitative, the leap of the evening is clearly toward the final love duet—as if the piece had been finished, in a *Turandot* turnabout, by the less ambitious but far more gifted Puccini.

Well, there's time enough for the profane later on if the sacred works out this well. One ought to quit this arena on a neutral note, something to recall the tradition rather than the break with it, for it is in France and Germany, not in Italy, that the radical experiments were made: *Cardillac* arrived simultaneously with the Puccini *Turandot, Antigone* and *Jonny Spielt Auf* with *Sly,* and *Die Dreigroschenoper* with *Fra Gherardo.* But Italian romantic opera had proved too viable to junk; refinements were preferred to total revision, and it is perhaps typical of the slowness with which composers and librettists moved that the long outdated premiere of Arrigo Boito's *Nerone* in 1924—shortly after the author's death but over fifty years after his sole other score, *Mefistofele*—was greeted as one of the most exciting events of the age rather than an afterthought.

And in fact, *Nerone* was still ahead of its time long after "its time" had ceased to be. Boito's music does occasionally court tedium, but his libretto on the interaction of opposing forces of vice and piety in Nero's Rome is a real corker, and must account for much of the work's great success at its La Scala debut. Christians spar with phoney magicians in the shadow of the dolmens of grand opera, from hymns to God to corybantic excess, and surveying the pageant is none other than Nero, a dramatic tenor and psychologically one of the most complex creations of . . . the age? Ah, but art this precious, it seems, is of no era—how else to explain Boito's fifth act, of which the verses alone were set down? At the Scala performances, only the four acts that Boito had himself composed were heard, and the evening closed with an effective enough tableau but nothing like the mind-bending conclusion Boito had planned, for in his last act the matricide Nero—or it is Nero?—dons *kothurnos* and *onkos* to take the part of Orestes in a snatch of Aeschylus with what appears to be a chorus of Furies—but are they actors playing Furies or real Furies?—while outside the theatre, Rome burns.

Besides the obvious connotations of the Pirandellian "life is a stage-turn is a dream is life" Chinese puzzle play, this obverse apotheosis, a Doric column topped by a dishrag, turns the world on its head, making a paté of reason and a *gulyas* of morality. It is truly amazing that Boito got the idea for his *Nerone* in the 1860s, and more than lightly ironic that when it finally took stage its prodigious intellectual proposition was the only such in Italian opera, at most flirted with in *Sly*, while the extraordinary *Wozzeck*, whose source also predates 1900 (by about sixty years), came along only a year after *Nerone*, in Berlin.

Enjoy it, you romantics, as long as it lasts. It doesn't last long.

3. GERMANY

The German-speaking nations traded briefly in verismo, though they're rather staid peoples for all that, and of course they had their Wagnerian playout as well. Of those who trod the boards in Klingsor's shoes, Siegfried Wagner leads the parade, by right of primogeniture if not for the quality of his fairytale operas, such as *Der Bärenhäuter* (*The Lazybones,* 1899) and *An Allem Ist Hütchen Schuld* (*Hütchen Is Guilty of It All,* 1917). Engelbert Humperdinck wins highest praise in the arena, mainly for *Hänsel und Gretel,* back in 1893, but a second fairytale opera, *Königskinder* (Royal Children, 1910), on a much less prancy libretto by Ernst Rosmer, won some following in its day for the composer's masterful entwining of leitmotifs and story-hour profundity, including an unexpectedly *Parsifalisch* third-act prelude. Still, the prize for the most thoroughly imitative Wagnerian opus goes to August Bungert, a Hellenophile who planned to complement the *Ring* with an even larger cycle—two cycles, actually, *Das Ilias* (*Hektor und Achilleus; Achilleus Tod; Klytemnestra*), and *Die Odyssee* (*Kirke; Nausikaa; Odysseus Heimkehr; Odysseus Tod*), the whole seven operas to be enshrined under the rubric *Homerische Welt* (*Homeric World*).

Like any proper "musician of the future," Bungert penned his own librettos; the *Iliad* section was never completed, but the four parts of the *Odyssey* were pounded out in authentic *Ring* style. The first pages of *Odysseus Heimkehr* (*Odysseus' Homecoming*), for instance, presented the darkling landscape and brooding god familiar from Nibelungenland on the shore of the Aegean and in the person of Pallas Athena flying around with lance and shield and ruminating on the changeless rule of the gods:

> *Was da lebet und webet,*
> *Was da rastet und hastet,*
> *Im welten Dunkel,*
> *Im Lichtgefunkel,*
> *Im wallender Welle,*
> *In luftiger Helle:*
> *Alles ruht in der Götterhand.*

What lives and weaves,
What rests and hastens,
In the world's darkness,
In the plays of light,
In the wandering waves,
In airy brightness:
Everything rests in the hand of the god.

Bungert's tetralogy was not quite the thrice-told tale it ought to have been in the hands of someone so possessed, but one could not help feeling that one had after all been there already. Poor Bungert had hopes of commandeering a second Bayreuth for *Homerische Welt,* but his pleas for a *Festspielhaus* of his own were enthusiastically resisted on all sides, and *Die Odyssee* came to light step by step in Dresden, *Nausikaa* in 1901 and *Odysseus Tod* in 1903 (*Odysseus Heimkehr,* Part Three, was the strongest, and had materialized first, better foot forward, in 1896).

Of authentic Wagner work there was little besides Bungert, but of Wagner-influenced opera there was more than enough, even if this meant a moratorium, for the duration, on the Word. As in France, the neoclassicists were on the field by the twenties, but in the meantime the romantics prevailed, through-composition bequeathing them a boon of no small quantity. Eugen D'Albert made a case for German verismo in the once wildly successful *Tiefland* (*Lowlands,* 1903), a succession of musical numbers without the recitative, restless and fraught, as befits the tale of an innocent shepherd lured from the clean air of the highlands to the foul valley, where he is married off to the local landlord's whore. In the end, the shepherd captures the heart of the woman, kills the nasty landlord, and takes his wife away to the good life in the hills: a slender action, with much room for villainy, drastic narratives, and an attractive opening portrait of the highlands, breathing pure air in a melody of fourths and little circlets of thirds reminiscent of the Rhine music in the *Ring*.

Using D'Albert's melodic and harmonic framework, Erich Wolfgang Korngold forewent the violence to fashion romances as thin as baggies, albeit with a sumptuous, some say Straussian, charm. Korngold was still in his teens when his one-acts *Der Ring des Polycrates* and *Vera Violanta* floated to the surface on a double bill in 1916, and he established himself as a man of the hour . . . just . . . with *Die Tote Stadt* (*The Dead City*, 1920), a dream opera based on Georges Rodenbach's novel *Bruges la Morte* and a notably decadent piece of ceaseless chromatism, rotting, deliberately, with images of ever ancient Bruges, a necropolis of medieval torpor. The tenor hero, a widower, has sunk into mourning for his wife, and he cannot pull himself out of it until he has expunged her shade by murdering her look-alike—this is the opera with the duet, "Glück, das mir verblieb," that used to turn up inescapably as a solo encore piece on soprano recitals through the thirties.

Korngold made it too easy for critics to scoff at his work by closing his career scoring movies in Hollywood, for "movie music" was just the term they had been searching for to describe the meandering, kitschy dysentery of *Spätromantik*. It would have been more correct to assess late romanticism as a collapse of shape, for in riding through-composition to the final stops of the trolley, these composers lost sight of the rise and fall, the arch and incision of musical dramatization, letting endless melody seem, in default of anything else, too much of a good thing. But in both subject matter and melodic coinage they were unstinting; the titles alone entice—Paul Graener's *Don Juans Letztes Abenteuer* (*Don Juan's Last Adventure*, 1914) or Rudi Stephan's *Die*

Ersten Menschen (The First Men, 1914). And breathes there a man who can fail to respond to the seamless, passionate outgiving of Korngold's heroic Queen Heliane in *Das Wunder der Heliane (Heliane's Miracle,* 1927), even amidst the admittedly unyielding orchestral flow? Consort of a cruel tyrant in a loveless country, she undergoes a chaste communion with a young messiah who has been tossed into prison for preaching a doctrine of love. Brought to judgment for treason, Heliane defends herself in a stunning solo of poetic candor, rife with sacred intimation:

> *Nicht hat mich Lust meines Blutes*
> *Zu jenem Knaben getrieben,*
> *Doch sein Leid mit ihm getragen,*
> *Und bin in Schmerzen sein geworden.*
> *Und nun tötet mich.*

> Pleasure of my blood did not drive me
> To that boy,
> But I bore his sorrow with him,
> And in pain became his.
> And now kill me.

The bulk of this repertory, adored in its day, has vanished, and it is not difficult to see why: lush harmonics and polyphonic excess dulled the intellect, lulling the senses but ceding nothing of substance to quell the hunger. Too many of these pre- and post-war romantics lacked the clarity of D'Indy or Roussel, the dynamics of Puccini or Pizzetti. With the premiere of Bartók's extraordinary *Duke Bluebeard's Castle* two years past, Emil Nikolaus von Reznicek's ordinary *Ritter Blaubart (Knight Bluebeard,* 1920) was not even treading water; it was drowning.

One man, who determined to save romantic opera from both decadence and revolution, wrote his masterpiece to detour opera back to its prechromatic vigor. This was Hans Pfitzner, and he would have none of the phantasms of the Korngolds. A creator who conceived of a "romantic cantata" on the nature of the German soul, *Von Deutscher Seele,* and who pamphleteered a diatribe on "The New Aesthetic of Musical Impotence," was not the type to investigate the involutions of dead Belgian cities or the last ride of a Don Juan; for him, the crystalline facets of German legend—*Der Arme Heinrich* (*Poor Heinrich,* 1895), *Die Rose vom Liebesgarten* (*The Rose of the Garden of Love,* 1901), *Das Christelflein* (*The Little Elf of Christmas,* 1917), *Das Herz* (*The Heart,* 1931).

Pfitzner's was the sacred arena, embodying to the nth what Henry James called "the associational process," the communion of artist and public—yes, Gide's fourth unity, so unified in this case that non-Germans tend to find Pfitzner's oeuvre unapproachable. A latter-day meistersinger, Pfitzner was regarded with idolatry by his fellow conservatives, and the focus of his work was a colossal apostrophe to tradition, the "musical legend," *Palestrina* (1917), to his own libretto designed as a statement of purpose anent the past, present, and future of music.

Heavily diatonic and contrapuntal, *Palestrina* was a reform opera, manned, unlike those of Gluck and Wagner, by the rear guard. Symphonic act preludes of some length bring the audience back into Wagner territory, transparent chordings plead for a breather from deteriorating tonality, and the work's central premise leaves no doubt that, in Giovanni Pierluigi Palestrina's challenge to restore ecclesiastical polyphony in 1563, Pfitzner was hymning his own story.

And yet for so personal a work, there is no sense of braggadocio or confession, not even in the moving scene in which the ghosts of Palestrina's late wife and nine Renaissance composers plus a choir of angels goad the reluctant hero into the composition of the Missa Papae Marcelli, an episode of intimidation, discovery, and glorious release, the myth of artistic synthesis. The second act, covering the final deliberations of the Council of Trent, strikes the foreigner as tedious, but Pfitzner's dialogue is deftly turned, and the brief third act offers a telling moment in the appearance, at Palestrina's house, of Pope Pius IV. It is scarcely three pages of score, but of a profound message to the listener. "Prince of music for all time!" is the Pope's pronouncement. "The Pope's servant and son!" So, in his reactionary treatise, did Pfitzner show himself as intelligent of his times as any renegade classicist. Pfitzner, too, was a leader, an overturner, but in the opposite direction. "Forward," he might have cried—"to yesterday!"

Less concerned than the Latins with the correct setting of linguistic cadence, these Germans and Austrians of the *Spätromantik* drowned the Word in dreams, and comedy suffered. Where, other than in the two classic comedies of this era, *Der Rosenkavalier* and *Ariadne auf Naxos,* was the whiplash of wordplay, the crackpot ease of true commedia? One romantic comedy intrudes here, not for any tensile sharpness, but because it alone harnessed the Wagnerian coach for horseplay, and it, too, is by Richard Strauss, to the "song poem" of Ernst von Wolzogen, *Feuersnot* (*Lack of Fire,* 1901).

This is not, however, the Strauss beloved by the prudent devotee. A rank sex farce, a smoking-car story, this *Lack of Fire* were better entitled "Lights Out and Tights Off"—really—for that's exactly what occurs in it, and in exactly that tone. In old Munich, on the eve of the annual fire festival of the summer solstice, a necromancer named Kunrad steals a kiss from a local belle, Diemut, who takes revenge by hanging him outside her balcony in a basket. The mob's laughing-stock, he takes *his* revenge by magically extinguishing every smidgeon of flame in the town, refusing to restore light to Munich until the girl submits to him . . . and soon enough, the lights do flicker on.

The real revenge in *Feuersnot* is that of Strauss, who had suffered humiliation in Munich, his hometown, with the failure of *Guntram* (inspiration apparent) in 1894. Less Wagnerian than its predecessors, *Feuersnot* nonetheless refers at one point to a certain Master Reichhart, whom the populace drove away— whose wondrous originality they rejected—as the real-life populace of Munich drove away Wagner. Clearly, Strauss is to be taken as Kunrad, the successor of Reichhart/Wagner, and *Feuersnot* is Strauss' *Palestrina,* however irreverent. In one pun-filled passage Wolzogen alludes to Wagner ("darer"), Strauss ("battle"), and himself (wohl gezogen—"well conveyed"), and just in case anyone hasn't gotten the point, Strauss quotes from the *Ring* and *Guntram.*

Wolzogen's libretto, cocked in a rapscallion Bavarian dialect, reaches past the fire festivals of central Europe into their gist, the fertility ritual associated with the extinguishing of fire followed by a ceremony of sexual and incendiary renewal—though here, of course, the situation is played for spice rather than mythopoeia (the censor at Dresden, where *Feuersnot* had its premiere, must have felt totally defeated). At the finale, when Diemut and Kunrad smile down at the townspeople from her balcony, the rite accomplished, the crowd has this dainty to offer:

> Imma, Ursel, Lizabet,
> Alle Mädeln mögen Meth!
>
> (Every maiden likes her honey!)

This was Strauss in his second opera, affrighting the public with the bizarre, and more distinctly Straussian than Wagnerian. Shrieky sounds, the tumult of *Till Eulenspiegel,* a nasty silence when the lights go out, and a potpourri of folkish tunes give *Feuersnot* a vitality unusual for the time, and if it failed to catch on as a steady entry, Strauss' next offering, the even more bizarre and shrieky *Salome* (1905), immediately planted itself in the front herb-box, right down by the kitchen window where everyone could see. Under, over, and commingled with Hedwig Lachmann's literal translation of Oscar Wilde's play, the composer prepared an oily collage, hued in the colors of a pre-Raphaelite's nightmare. Some of the more distinctive motives, such as those identified with the prophet Jokanaan and Salome's passion for him, are far removed from the perfect fifth of the Flying Dutchman or the plastic *Ring* shapes that the modern leitmotif was weaned on; Strauss' thematic reference table is less clearly derived than Wagner's, less tautly connective, less elastic. In Schoenberg's *Gurrelieder,* written before and performed after *Salome,* one motif grows into another to build the whole piece out of one ur-idea, but Strauss is more interested in the neuropathic textile of his piece, in establishing a point of view for the corruption of Herod's court—corrupt Tetrarch, corrupt consort, corrupt stepdaughter . . . even corrupt saint, to judge from the augmented fourth of his motif. Or are we meant to view it all through the eyes of the perverse heroine?

More than a few critics held that *Salome* was taking mannerism too far, but this was after all the trajectory of the fin-de-siècle, to scourge innocence with the grotesque (and the grotesque would permeate Strauss' operas through the 1920s). *Gurrelieder,* too, joined the conspiracy, in its "Wild Hunt of the Summer Wind," an almost Aristophanic gust of frenzy, but demented rather than cranky. The innocents can easily avoid the seldom-performed *Gurrelieder,* but *Salome* is everywhere, and if constant repetitions have exposed the cheesiness of its weaker pages, it has held up on the simple power of its ambience and the corrupt luster of the famous final scene. It does entice, with the curtain shooting up on the famous clarinet run, with that disgusting repeated note during the beheading—four double bases pinching a B Flat at the topmost shrill of their compass— and the obnoxiously exotic dance of the seven veils. But how could a veil dance be tasteful?

Elektra (1909), Strauss' next work, like Salome a one-act, marks the commencement of the composer's partnership with Hugo von Hofmannsthal, a perfect lesson in how much more comes of genius when it is ignited by genius. One must call it piquant that so gifted a writer as Hofmannsthal chose the "loudest" orchestra then going in which to hearse his theatre, yet on the other hand Strauss was the most distinguished composer of Hofmannsthal's timeplace: who else, who better? Theirs was never a meeting of kindred spirits, and differences of education and temperament kept the earthbound Strauss from understanding the fullness of his partner's librettos (more than once he mistook stage directions for lyrics and set them accordingly), but theirs was the most fertile collaboration since Mozart and Da Ponte, begetting at least four masterpieces of musical theatre— and such music! What's this about an earthbound Strauss? The man who composed the A Flat Major section of the recognition scene in *Elektra,* "Es rührt sich niemand"? Or the peculiar chain of chords in *Der Rosenkavalier*'s presentation of the rose, like a bevy of undiscovered gems? Or Mandryka's joyful "Mein sind die Wälder" in *Arabella,* as ample an outpouring as the estates that he describes and which breaks out with such eclat at the final curtain as the heroine dashes up the stairs on the last night before her wedding? Earthbound, this Strauss? A bad match for the consequential Hofmannsthal? No, who better?

With its giant orchestra (well over a hundred players), *Elektra* went even further than *Salome* had into the cacophony of obsession, but with Hofmannsthal for a librettist in place of Wilde, this was a vastly different operation, a monumental nexus of revenge tragedy, psychological study, and classical reinvestment—the regeneration of old themes via modern interrogation. Where *Salome* whined and went limp, a mauve caprice, *Elektra* is a coiled spring of inevitability. They are about equal in length, but how much more is packed into *Elektra's* late afternoon than *Salome's* night: first, the heroine's monologue, a case for Freud not merely in word-pictures, but in sounds as well, insatiable natterings, outbursts, screaming; Elektra's confrontations with her sister and her mother, the latter scene presaging the expressionism of atonal opera; the recognition, which opens, so tellingly, as destiny moves Orest into position, with his sister's unwitting question, "What do you want, stranger?," and her failure, after years of rehearsing the ultimate act, to be ready with the axe when the time comes. On such irony did the Greeks believe a tragic theatre must depend, and of such families as this of Mycenae did the drama festival tell. Here is the opera that the Camerata could not face up to, and did not invent.

Ernestine Schumann-Heink faced up to the part of Klytämnestra, to her eternal regret, and others were of her mind in thinking the work and its authors fit for the nuthouse—they dubbed Chrysothemis' "Ich hab's wie Feuer in der Brust" the "Chrysothemis Waltz," and said the music smelled. Such hydrophobia overwhelmed them and, one must concede, overwhelmed the Word as well; this is romantic opera bursting into, not out of, Bedlam, nothing like neoclassical recovery. This is what the romantics taught: that there is no exit, no deus-ex-machina, no convenient doom. Once Orest enters the palace to murder his mother and her lover, he never reappears, though he is urgently summoned. It is as if, having fulfilled his function, he has quit the stage of life, while Elektra collapses, her work done, and Chrysothemis, bewildered by the gaps that have closed up so suddenly around her, batters helplessly on the great doors of the empty palace and the shade of Agamemnon savages the place.

But *Elektra* lunged as far in that direction as Strauss and Hofmannsthal needed to go. For Strauss especially, the way onward was to grow increasingly sunny, and for their next project, a "comedy for music," the pair engaged Vienna in the first years of Maria Theresa's reign, raising the curtain on the morning after an ardent tryst and drawing it to on the remains of a concupiscent rendezvous in a low tavern. This is, of course, *Der Rosenkavalier* (*The Cavalier of the Rose,* 1911), the "waltz opera." "But how dare Strauss write waltzes for a setting several decades antecedent of the waltz?" There is always something to grumble about, if one is resolved to grumble, and as far as that goes, the setting also predates *Spätromantik* harmonies and the "psychological polyphony," as Strauss termed it, without which any opera of the 1900s would have been a pale parcel indeed. What did they want? Ländlers? Sarabandes? Opera wants its atmosphere—interpretation, not reconstruction—and whether or not the Waltz is suitable for Vienna in the late eighteenth century, it is eminently right for the *character* of Hofmannsthal's etchings. *Der Rosenkavalier* is a light comedy, of a sentimental rapport for all its lumpen slapstick, and the most unprofane of comedies. *Die Meistersinger* has more nobility and *Falstaff* more humanity, if one *will* compare, but *Der Rosenkavalier* is the intimate one; it touches most closely, with no less its share of sincerity, and one is touched by it always. Let it have its waltzes.

Of a sort, the work was by way of homage to *Le Nozze di Figaro*, adopting from the earlier piece the trouser-role adolescent, hints of class-war friction, lonelyish older woman, lack of a principal tenor, and attempted seduction that doesn't quite work out. These are incidental blandishments, and might be dismissed but for the fact that Strauss openly claimed an allegiance with Mozart, a tantalizing opportunity for his detractors. Said Cecil Gray, "the divinely innocent and virginal Mozartean muse cannot be wooed and won like an *Elektra* or a *Salome*. All we find in *Der Rosenkavalier* is a worn-out, dissipated demi-mondaine, with powdered face, rouged lips, false hair, and a hideous leer. Strauss' muse has lost her chastity." Further, one could score the composer for overstating his psychological polyphony; much of Hofmannsthal's text is rent in the cyclone, and some moments of the long score pitched at an emotional level better suited to affairs of love and death than to those of love alone. Having got that out of the way, we can repudiate Gray and his happily up-shown sensitivity, for much of the vocal-orchestral ambience of *Der Rosenkavalier* does truly radiate a Mozartean delicacy, as in the Marschallin's Act I monologue, in the Sophie-Octavian duet *con duenna* after the presentation of the rose, or in the stainless songfulness of the last lyric, "Ist ein Traum, kann nicht wirklich sein, dass wir zwei beieinander sein"—who would have anticipated anything that fresh to come of a simple G Major-D seventh-e minor-D seventh progression?

On the way to neoclassicism—to parody, control, and to a reinstatement of the tension of structure—one might remark the famous aria of *Der Rosenkavalier*'s nameless Tenor during the Marschallin's levée, "Di rigori armato," a cordial entailment on the baroque performed with a romantic's flair of exorbitance without embellishment. For all its Italian temper, it is not at all foreign to Strauss' regular sound (unlike certain moments of Italian and English quotation in *Die Schweigsame Frau,* a classical piece from an older and somewhat classical Strauss); indeed, the entire levée scene is a very gentle brand of re-creative pastiche, almost but not quite a viva-voce plunge into Hogarth. The very model of romantic comedy, *Der Rosenkavalier* stands at the opposite pole to modern comedy, that nightmare world wherein awful clones gaze out of mirrors. No, *Der Rosenkavalier*'s mirror reflects a pretty world not without its momentary dismay—this, too, is Mozartean—but while the young lovers, Sophie and Octavian, achieve a splendid finish, the Marschallin vanishes in half-light neither happy nor truly tragic, and this yields a pathos not of Mozart's era.

But not exactly of ours, either. Hofmannsthal's wisdom led him into the fertile territory of new-old in almost everything he did. As with *Ariadne auf Naxos,* but not so sharply, he penned *Der Rosenkavalier* to prove old ways in new means, yet here his vision is less extreme at its wings; not even the garish ruination of Ochs' dinner with "Mariandel" in Act III has the blunt staccato of Zerbinetta and her crew, and there is likewise nothing in the older work to rival the character of Ariadne, the woman who wills the god. But even so, the same double-bias of back-alley buffo and noble proportion infects the mood of *Der Rosenkavalier* as it later would that of *Ariadne,* if in a less quixotic temperament. Act I, in the Marschallin's bedroom, is lush, naughty, and introspective, neatly halved by the particulars of the levée, a "genre painting," a duet with trio and inter-

lude—one thinks of Rubens (dimly, however)—perhaps a *conte*;
Act II, in the main hall of Von Faninal's town house, is more
extroverted, exciting and then almost pushy, a stunning portrait
for the Presentation of the Rose (Lancret, possibly?), and a detail from Breughel for Ochs' waltz scene—but let Congreve carve
the motto ("May such malicious fops this fortune find . . .");
Act III, in a low suburban tavern, is farce full out, a mezzotint
entitled "The Rendezvous," a *fête galante* off on a tear, and at
last that joyous trio and duet—all this in a romantic comedy, a
"comedy for music." Clearly, there is much more to comedy now
than there has been ever before.

At the time of *Der Rosenkavalier*'s premiere at Dresden,
Strauss was not unlike Massenet in that his issue was eagerly
awaited by the international music community; that was all the
two had in common. Massenet, as the years passed, ascended no
plateaus in form or content, while Strauss, even without Hofmannsthal, submerged himself in ontogeny, seeking a better
way than his psychological polyphony without losing the psychology or the polyphony. Also, Massenet never won the confidence of a stage director of the brilliance of Max Reinhardt, who
lent much needed assistance in the mounting of that first *Rosenkavalier*, particularly in his plastic blending of Fragonard and
Hogarth for the stage pictures. The set and costume designs of
Alfred Roller, still talked of and reproduced today as a wonder
of the age, seconded the elegance of the authors' idyll, and the
touches of Viennese argot ("I weiss halt nit, ob i dös derf")
proved no less eye-opening. Other productions followed in short
order—at Nürnberg one night after the Dresden opening—and
a new work was permanently established. Has it happened since?

Ariadne auf Naxos, we know, retains an infirm hold on the
repertory, and the next Strauss-Hofmannsthal partnership, *Die*

Frau Ohne Schatten (*The Woman Without a Shadow,* 1919)
remains one of the most contested items among the cognoscenti,
for some the finest bloom of the sacred garden, for others an offensive weed swaddled in hothouse orientalia and symbolism.
Not much of a cognoscento himself, Strauss approached the
project reluctantly; this was Hofmannsthal's hobby horse, and
the libretto does have one weakness in that the story-beneath-the-
story is one of those better suited to pure literature than to the
temporal confines of opera. Indeed, *Die Frau Ohne Schatten*
ended up a lengthy work that can only be grasped after one has
read Hofmannsthal's inadvertent gloss, a short novel penned in
conjunction with the text that covers more fully much of what is
neglected in the libretto.

As *Der Rosenkavalier* stands to *Le Nozze di Figaro,* so does
this *Frau* to *Die Zauberflöte,* dominated by a Sarastro figure,
Keikobad (who never appears, however, but who tyrannizes the
action through an omnipresent and threatening three-note
theme), and deploying couples from both the masses and the
classes in trials of purification. Here, however, the test is of a
more interior challenge than that met by Tamino and Pamina:
to pass in this work, one must overcome the self to comprehend
the essentiality of human bonding. Schikaneder's elemental
gauntlets and Mozart's flute are developed by *Frau*'s authors,
amidst the farouche confluences of the fairytale bazaar, into a
mythic harrowing difficult to cast and almost impossible to stage
competently—and yet, as almost every revival has proved, it's
worth the extra effort.

The woman of the title, called the Empress, is the focus of the
piece, hers the most strategic trial, for she is no real woman at
all, but a fairy, immortal and infertile. At the behest of Keikobad, her father and the ruler of the spirit kingdom, her human

husband will turn to stone in three days if she does not locate a
shadow (fertility) somewhere, and it is while in the world of
humankind, seeking one, that she senses the self-fulfillment de-
nied her spirit situation. Ultimately, the Empress is presented
with the sight of her husband, petrified but for his two glowing
eyes. She can save him by stealing the shadow of a mortal
woman, but her last three days among mankind have been
crowded with event, and she has developed, if not a shadow, an
ethic, and she refuses the bargain. At that Keikobad pardons all
in a happy apotheosis, for his daughter has thus earned not the
surface identity of humanity—the shadow—but its true shape,
self-conquest.

It is too bad that Hofmannsthal's prose *Erzählung* on the same
tale isn't better known, for it accommodates the all-important
alternation of the human and spirit worlds that can be lost on
stage, along with such details as the value of the Empress' magic
talisman, the development of her affection for the people she
encounters in the "real" world, and the silhouettes of the Unborn
Children, supplicating offstage voices in the libretto, but busy
characters in the story, who in one scene treat their own yet-to-
be father, the Emperor, to a banquet somewhere in Keikobad's
domain.

As with *Ariadne,* the bourgeois Strauss who was reportedly
out of his depth in his Hofmannsthal collaboration voiced the
penetralia of the narrative with no little acuity, in a system of
leitmotifs that match the poet's intellect with lightning flashes
of their own. The work is obsessed with rising and falling sevenths, giving it an ethos distinct from Strauss' other operas despite the indelible signature, and the use of the violin in certain
passages seems a ready syndication with the expressionistic Viennese school launched by Schoenberg. The fourth scene of Act II

offers such a moment; in its niche between waking and dreaming, the human and the spirit, exposition and climax, it takes the perfumed exotica of legend into the more telling communication of myth. In this short scene the Empress is found envisioning her husband's progress into Keikobad's granite net while the voice of a falcon cries, "the woman casts no shadow—the Emperor must turn to stone!" Starting awake, the Empress realizes the guilt of her incomplete being—that much is Hofmannsthal—as the orchestra, retrieving a host of spirit and human motives, pounds out one in particular, four heavy chromatic chords associated with the bargain she has made for a shadow. But what the Empress must learn is that she can only "grow" her own, not buy another's, and as her nightmare builds, the bargain motive is thudded into meaninglessness, a useless expedient—and *that* much is Strauss.

This apogee of the sacred landed, just after the war, in a shattered Europe whose artists soon seemed to be fighting any attempt at moral recovery with corruption in literature, the theatre, and music. Stravinsky had detonated *Le Sacre du Printemps* and Schoenberg his *Pierrot Lunaire* just before the war, and if anything the hostilities inspired the front lines of the art world with statements apter of disease than of life. But so they saw life, and just as Marcel Duchamp's Nude Descended its Staircase, Stravinsky and Schoenberg, too, made their apparent descents, the one into jabbering unmelody, the other into filth and innuendo. Or so the public thought.

What, possibly, was to have happened just then—and would have if the muses hadn't been playing ducks and drakes with the diehard romantics—was a proper neoclassical revival, with its amenity and contentment, but the agony of modern art rejects the discretion of form except for use in burlesque and seedy pastiche. *The Waste Land* was universal to the continent, and permitted no easy way out of romanticism, no clear substitute save parodistic commentary. This is what satire is for: it elbows romance aside, disenfranchises the hugeness and the fantasy, sneers at singers. Satire is words.

So, in 1924, came Strauss' Word piece, *Intermezzo*, no companion for Eliot's poem, true, but one way of dealing with the Word without turning "decadent." *Intermezzo* is Strauss' most underrated work, and his entirely, as it happens, for Hofmannsthal would have nothing to do with his partner's intention of dramatizing a quarrel he had had with Frau Strauss, a former opera singer (under the name Pauline de Ahna) and a notorious termagant. The terror of first-night green rooms, she had been *Die Frau Ohne Schatten*'s biggest detractor, thinking it such rubbish that she refused to be seen walking back to the hotel in Strauss' company the night of the premiere in Vienna. Indeed, how could the cultivated Hofmannsthal find the possibilities in an opera about a shrewish wife abusing her husband on a mistaken assumption of infidelity, amidst scenes of disconcertingly humdrum domesticity, making no attempt to hide the autobiographical nature of the piece—nay, even trumpeting it? Disgraceful. Humiliating. Only a hopeless *Bürger* like Strauss would attempt such . . . such wanton naturalism.

Intermezzo, a "bourgeois comedy with symphonic interludes," has the lilt and tenderness expected of Strauss, but a tangy *Wortspiel* as well, less charged than the prologue of *Ariadne auf Naxos* and unblushingly domestic—an *intermedio*. Natural it certainly is, providing an unappetizing portrait of the composer's difficult wife, but if Pauline didn't mind, why should we? Far from taking umbrage at her operatic counterpart, she even told Lotte Lehmann, the original "Frau Storch," not to be stingy with the temper. Still, even she must have been taken aback by her husband's first-act finale, in which she tearfully disturbs her sleeping child with tales of his father's infamy, getting nothing but abuse herself from the infant, loyal to papa. Critics have decried this breach of taste, but in performance it comes off as a comical moment, one of reverse charm and somehow, withal, absurdly touching.

Intermezzo, a longish comic intermezzo hyphenated by musical intermezzi, brought the chimes of the front hall clock into the opera theatre (to balance the plebian flat of *Louise*), and some of the stuff of life—dining-room spats, idle letter-writing, a sleighing mishap, screaming at the servants, and the famous card-playing scene, a scat game. All of this unfolds in a cunning tour of operatic expression, from bits of spoken dialogue and dry recitative to arioso and, of course, the rich symphonic intervals. In a foreword to the published score, Strauss took the trouble to expound the aesthetic he had evolved for *Intermezzo,* noting especially the question that damned composers and librettists doubtless spend their eternity of hell trying to settle—sing it or speak it? Strauss had no intention of arbitrating the dispute (not yet—he did find an answer in *Intermezzo*'s second cousin, *Capriccio*), for in 1924 there was no answer.

German opera fulminated with possible answers, and from men without Strauss' attraction to Mozartean and romantic perpetration. There were Dionysiacs in the earth in those days, aware of Apollo's verbal propriety but contemptuous of his grace. What they had in mind was operas about black jazz musicians who conquer the world, or about the rooted vanities of humankind as viewed in the underworld of London, or, sometimes, about total nonsense. One couldn't expect the grand line from them, oh no; more often than not they perforated their own designs with the most intrusive musical framework, thrusting fox trots on their public. How, possibly, was one to concentrate on an opera if the music kept making one think of dance halls and cabarets?

But that, it seems, was their point exactly: think, spectators, of dance halls and cabarets, not of romance. Think rather than enjoy. If possible, do both—but do not abandon yourself to sensuality. This can only be a satirist speaking, and here come the satirists, the German race of which stands closer to Swift than does Eric Satie or Ermanno Wolf-Ferrari. In German opera, we contact the ferocious moralists, Ernst Křenek and Bertolt Brecht and Kurt Weill, who make snowballs of sociopolitical drama, ice them with gleanings from sleazy pop music, and roll them down the hill to collide with modernity, with jazz and atonalism. In a way, they are on Arnold Schoenberg's team, or he on theirs, for in the employment of vastly different compositional and thematic techniques, Křenek, Weill-Brecht, and Schoenberg all pressed the profane upon Europe's operagoers. They made quite an impression.

So. It is classicism again, "up to a point, Lord Copper," only up to a point. Romantic discovery will not cede the front all the way down the line, for too much had been discovered to be put by, but an outgrowth of rigorous forms, lighter orchestration, reduced forces on stage, and, of course, the word-charged score now clearly identifies itself in Germany. Such extremist musicians as Křenek, Weill, Schoenberg, and Alban Berg merit their own chapter; for now, let us view their more moderate coevals to see what, if anything at all, remains of the gallant Camerata.

Of the pseudomythic subjects beloved in the seventeenth century, nothing survived in German opera. Greek settings and characters had come back into fashion, yes—Richard Strauss alone has already given us Elektra and Ariadne and will investigate Helen, Daphne, and Danae later in these pages—but the "natural" harmony of the simplistic worldview has faded. The natives, restless, have turned to comedy. For German-speaking artists, the absurdist unrealism of Jarry and Pirandello proved irresistible, especially for the musical expressionists, though their work was largely kept under wraps, for knowledge is death, by timid impresarios (Schoenberg's *Erwartung*, for example, was composed in 1909 and published seven years later, but not staged until 1924).

Emil Nikolaus von Reznicek, among others, latched onto the wacky new commedia, in much better humor than the impudent Krenek and Weill. In his one-act *Spiel Oder Ernst?* (*Play or Reality?*, 1930), libretto by Poul Knudsen, Reznicek borrowed morsels of *Rigoletto, Don Giovanni, Die Fledermaus,* and *Tannhäuser,* plus a few bleeding chunks of Rossini's *Otello,* to accompany a trio of singers and their pianist in a rehearsal of the Rossini. Otello, Desdemona, and Brabantio taste of the *Pagliacci* syndrome—is it an act or life?—are visited by a strange bobby-soxer who worships tenors, and eventually accuse the accompanist of making trouble and chase him out of the theatre. Overblown cadenzas demolish the bel canto epoch, and Knudsen even turns his own plot premise into a wacky ruse, for this was what was being done, especially in one-act operas, something of a rage at the time. Paul Hindemith made a vocation of the short piece on his way to severe neoclassicism, as in a parodistic puppet play, *Das Nusch-Nuschi* (1921), or in a theme-and-variations affair called *Sancta Susanna* (1922), or in a violent domestic farce, *Hin und Zurück* (*There and Back,* 1927), which reverses its action (but not its music) at the halfway mark.

Krenek, too, turned to the mindless romp in *Der Sprung Über*

Den Schatten (*The Leap Over the Shadow*, 1924), to his own libretto, a Dada society farce plumped up by a detective and a spiritualist and set in Křenek's perverse autograph, which at that time was shifting from the atonalism of his three symphonies and four string quartets to the *barbarisch* climate of the dance crazes. *Jonny Spielt Auf*, to be heard from in the next chapter, is the work by which he is best known, or, more exactly, most forgotten, but like Mascagni he has much to offer the wayfarer in the way of refreshment.

Too intellectual to win the mob's affection, Křenek patrols the archives with an amazingly various output, everything from the flimsy one-act to the gargantuan historical pageant, and all wrought of a stubbornly individual congeries of technique that attracted admiration instead of appreciation. Even his fox trots had craft, as if the beady eye of the tone row were peeping through the drag. His specialty was the *Zeitoper*, the music drama that incorporates the news of the day . . . fad opera, one might say, fetish play. In the mode, Křenek applied the claw hammer to austerity in *Leben des Orest* (*Life of Orestes*, 1930), an uproarious dig at the House of Atreus and a profane complement to Strauss' *Elektra*. A work of many moods, it discouraged anyone who would empathize with the wandering protagonist, snapping up like a jack-in-the-box whenever the situation threatened to turn affecting.

While the satiric classicists depraved opera, the serious classicists stripped it down and forced form upon it to rediscover the structural bases looted by the *Spätromantik*. Essentially, however, both occupied wings of the same plane in a flight of self-conscious testimony. Křenek might slip into a tango, Paul Hindemith into a passacaglia; either way, one couldn't help but notice. One pulled back, his attention decoyed. One heard the words and music as if from afar, or up too close, or vertically, the way one gauges a graph line. But never could one dive in and float. The mystery is dispelled.

France enjoyed *Padmâvatî* and *Antar*, Leoš Janáček's *Káta Kabanová* was heard at Cologne, Italy embraced Pizzeti and Zandonai . . . enter Hindemith's *Cardillac* (1926), a historic midpoint for neoclassical opera. From a story of E.T.A. Hoffmann about a jeweler so enrapt with his own artistry that he murders his clients to recover his gems, Ferdinand Lion provided Hindemith with a gripping, provocative drama; the standard line holds that Hindemith traduced Lion's theatricality with contrapuntal formalism. So prevalent is this view that Hindemith felt compelled to romanticize the piece in 1952—to share, unlike his goldsmith hero, his work with humanity. But the original version, which is still heard in revivals, is the better of the two, and its sharply delineated methodism, once one acclimatizes oneself to it, augments rather than diminishes the experience.

For example, the opera opens and closes with extended choruses, the first stormy and discrepant, a series of tiny staccato figures, and the final soothing and richly tonal in long arches of 9/4 metre. Thus the people of Paris frame the action, outraged at the commencement (by the rise in the murder rate) and placated at the finish (having killed the murderer), and in between each scene has its formulation, its form—unusual for a psychological thriller, yes, but not unthrilling. What could be more chilling than the famous flute duet in the mimed love scene of the knight and the lady, when we know that Cardillac is going to slip in at any moment and repossess the gold belt he sold the knight? Critics have commented on this metal nocturne as being deficient in passion, but onstage it never fails to raise the gooseflesh, not least because the dagger blow, the knight's bloody tumble, and the lady's scream of terror strike in total silence.

From the two pianos and six wind instruments of *Hin und Zurück,* one year after *Cardillac,* Hindemith graduated himself to two operas of overwhelming scale shortly before and then a decade after World War II, but the ultimate in grand classical opera had already been attained by Ferrucio Busoni, the emissary of the "dangerous future" that Hans Pfitzner preached against. Busoni's summum bonum was *Doktor Faust* (1925), an undertaking of fierce metaphysical pretension, fully realized, for which in a way Busoni's earlier work might be seen as practice pieces. All Busoni is divided into four parts, piano virtuoso, theorist, composer, and opera librettist, any one of which marks him as a man of stimulating procedure. Perhaps he more than anyone led the way from romanticism into neoclassicism, and in his words no less than his music.

Modern classicism rebuilds the present on the discoveries of the past, and as Busoni's *Turandot* in 1917 reengaged the world of Carlo Gozzi, its partner on the same bill, Busoni's *Arlecchino,* delved more variously into old art, quoting Dante, Mozart, and Beethoven while the avid characters of the *commedia dell'arte* race around, fall in a heap, pick themselves up, and do it again. The setting is eighteenth-century Bergamo, the natural habitat of Harlequin, a speaking part and the mastermind who gulls Dottore Bombasto and Abbate Conspicuo, cuckolds the tailor Ser Matteo del Sarto, and deserts Colombina for Matteo's wife. Laid out in four little movements—Harlequin as rogue, warrior, husband, and victor—the score pokes ceaseless fun at Italian opera, from Pergolesi to Verdi, and borrows the soldier's march from *Fidelio* for the second movement, transforming it into the trio section of a heavenly finale when the actors prance out for their bows. This is modern art, adapting the past, and of course Arlecchino delivers the envoi: "Does not everything repeat itself, and always in the same circle?"

Moving into the innermost circles of life's hell—and still setting his own librettos—Busoni then turned in *Doktor Faust* to celebrate man's quest for self-knowledge and the "ewiger Wille," the eternal will that transcends mortal death. Ah, a Wagnerian concept, with the fervid "it shall be" of the synthesizer—but in fact a quite inclusively romantic theme, from Goethe to Beethoven to Nietzsche, and, anyway, Busoni learned nothing from Wagner. His gods were Bach, Mozart, and Liszt, and as with *Turandot* and *Arlecchino*, in *Doktor Faust* Busoni found himself, like the two-headed Roman god Janus, half disposed to review the past and half compelled to arrogate the future.

Left unfinished at Busoni's death in 1924, *Doktor Faust* was completed by his amanuensis Philipp Jarnach and quickly mounted at Dresden a scant year later, so great was the interest in hearing the composer's sole large-scale work. *Cardillac* was still a year off, and *Wozzeck* would not be heard till the tail end of 1925, so the discrete shapes that Busoni applied to the sections of his drama as Hindemith and Berg did to theirs forced on one the difficult novelty of having to experience them as sections, of having to receive the narrative through the artillery of German symphony. Faust's invocation of six diabolic spirits calls for a theme and variations (as had, incidentally, Ariane's unlocking of her six permitted doors and one forbidden door in Dukas' opera, an early example of the luring of Wagnerian program back into its classical roots), the church scene in which Busoni's equivalent of Valentin is tricked into confessing to Mephistopheles and then murdered takes the form of a rondo, and a masque at the court of Parma assumes its outline from a suite of dances.

Classicism? Yes, but of a very romanticized kind, not only because of the subject matter, but because of the searching nonfinite borders of this at-first-glance heavily bordered work. The clear frontiers that separate, that classify, the parts of the whole, give way to the horizon at the edges of the evening, when Busoni himself, styled "the Poet," must step before the curtain to deliver at first a lengthy prologue and at last a brief valedictory, both in heroic couplets. Don Juan and Merlin also he had considered for his opus, he explains, for like Faust they bring us closest to myth, to Western absolutes, and for Busoni the sacred absolutely called for fantasy, that of custodial alchemists like himself:

> *Drum hielt ich Umschau unter allen jenen,*
> *Die mit dem Wunder wirkten, Hand in Hand:*
> *Ob gut, ob böse, ob verdammt, ob selig,*
> *Sie ziehn mich an mit Macht uneiderstehlich.*

> And so I made a survey of them all
> Who worked with magic, hand in hand:
> For, good or evil, damned or blessed,
> They draw me on with irresistible might.

At last he chose Faust for his hero, the most sempiternal of the three. His source was not Goethe, but an old puppet play such as was seen before the gates of many a medieval city, and the sport of Punch and Judy is here assigned to the Devil, who disguised as a court chaplain piously gives the Duke of Parma his benediction with a claw, who shows up in the tavern scene to heave a dead baby at Faust, and who, at the final curtain, now in the clothes of a nightwatchman, notes Faust's corpse and says, "Has something unfortunate befallen this chap?"

A comedy? Yes, but only in the sense that its foundation lies in farcical revue, and in its circular, no-rise-no-fall structure—and yet, everything does not, as Arlecchino opined, repeat itself always, not here. *Doktor Faust* is the antithesis of *Palestrina,* just as Busoni was that antipope who excommunicated the *Spätro-mantik*. Pfitzner's hero seeks self-awareness in tradition, Busoni's in both tradition and the unknown on the nearest limits of the mystery, and the spoken epilogue consolidates Busoni's idea that tradition leads to new traditions. The circle closes with new arcs subsumed:

> *Das gibt den Sinn dem fortgesetzen Steigen—*
> *Zum vollen Kreise schliesst sich dann der Reigen.*
>
> (This is the meaning of the preceding graduations—
> The round dance closes itself in a full circle.)

Inexpressive, they called it, obdurate. But no expressive, yielding work of that time equals this one for outgoing invention—in the tympani figure representing Faust's enemies banging on the door; in the Devil's sarcastic high notes, the B's and C's; in the tremendous Easter chorus that accompanies Faust's signing of the pact; or in the famous shouting match between Catholic and Protestant students in the tavern, Te Deum versus "Ein' Feste Burg." True, it is a difficult piece, even with the holiday of the comic puppets scrounging about in the intellectual quest—but as Maeterlinck's Ariane explains, "What is permitted teaches us nothing." It was said of Busoni's "Fantasia Contrappuntistica," a piano piece with the prejudices of Bach and the pride of Liszt, that of all the day's keyboard virtuosos only the composer could hit the right balance of the utter and the conniving called for, and if even his fellow musicians couldn't keep up with Busoni, the untutored public was totally out of the running. Busoni wanted to conduct this public into the dangerous future, Pfitzner wanted to soothe it, and neither one convinced it as surely as did Richard Strauss, who simply held to the middleground. Even today, Busoni is a figure of legend, a weird whose future, now that it's the present, never came true.

IV

Jazz, Fugue, Tone Row, and Other Wonderful Motions

"In America, the land of the free, once lived a nigger boy who danced far and stepped near and always danced in style," sings the crowd, edging into a shimmy in Ernst Křenek's *Der Sprung Über Den Schatten*. Freakish polytonal blocks of chords in parallel motion accompany Orestes and Electra in Darius Milhaud's *L'Orestie*, the singers, unaccustomed to fighting the orchestra, looking a bit startled—wouldn't you? Conservatives decry the decay they hear in Alban Berg's adaptation of Georg Büchner's play *Woyzeck*, but they are drawn back for a rehearing, and another, and at length they are won. It is jazz, and an abrupt neoclassicism, busy with shapes more than melody, and atonalism, invented so music could be expressionist, too. It is opera gone mad.

No one term should be stretched to cover all the developments represented above in synecdoche, but they did all occur at about the same time—roughly from 1910 to 1935—and they do have this in common, that each delivered a revaluation of the Word quantum in music drama.

Let us call them the operas of contraposition.

And this being a chapter on contraposition, it must be outrageous, for this is an outrageous era to cover, of ramshackle discipline and cold-hearted intent—or so thought the outsiders. To them, it meant the end of opera as they knew it and as far as they cared to, and to others, the beginning. Whether or not the fears were justified, the operatic harvest does get leaner after 1938 (a good year for German opera, though—one wouldn't have thought it likely, given the political situation—including the premieres of *Daphne, Mathis der Maler,* and *Karl V*), and except for a bumper crop of new titles immediately following the armistice in 1945, the soil yields small shares thereafter.

But if opera is in weak health at the moment, contrapositional technique—the playing off of music against words, the "antiexpressive" language of the twenties and thirties—didn't hurt it. On the contrary, the razzing Křenek and Weill, the precisionist Stravinsky and Milhaud, and the idiopathic Schoenberg and Berg brought vitality to the musical theatre and, frankly, no one was ever more "expressive" a theatre composer than Alban Berg in his two masterpieces, *Wozzeck* and *Lulu.* The emergence of them all in their distinct guises covers the territory of neoclassicism, that only partial answer to the questions left unanswered by the romantics, and the Word glowed, seriocomedy took stage, and discrete forms divided the sum into its parts. One noted the devices, sensing if not reading the overall scheme, and the profane took a holiday, dressed to kill in the nudity of text. The Word is all to the profane, the preciseness of action that, in musical terms, overturns the spellbound through-composition of late romanticism for a through-composed alertness, the Word in notes. This is why audiences have such difficulty "getting" atonal opera at first: it plays tragedy in the rhythm of comedy: its tunes are sung in words.

This is what the neoclassical revival did to romance, as we shall be seeing shortly. "Program" is the tension of romantic opera, the Dionysian approximation, and the best way to illustrate this is to pinpoint opera on the verge of revolution in an oft-quoted story: leading an orchestra rehearsal of his *Iphigénie en*

Tauride, Gluck found the string section balking at Orestes' line, "Le calme rentre dans mon coeur," set as it was to an obsessive syncopation in the violas. The players thought this an absurdly inapposite accompaniment for "The calm reenters my heart," and in that sensitive age of the Word, they could not bring themselves to proceed with the movement. "He's lying! He killed his mother!" the composer replied. "Keep playing!"

In classicism, however, the tension is "system," the Apollonian exactness, the tempo of comedy. Romanticism tears at shape; classicism recovers it, for it is the more agile of the two, and the more rigorous. Oh, the hubris of comedy!—quicksilver, ridiculous, and natural, although this is a nature that the Camerata would have had trouble placing. Tragedy is there to reveal the myths according to the lights of each newborn age, but comedy buys us time till the Lord come again. Comedy is the Word intransigent, unmitigated by the efforts of Orpheus. It doesn't ask one to feel, but to be. And when comic lay is given a romantic subject to develop, the result can be disconcerting. This is contraposition, the polity of the profane.

Nothing could have been more profane than *Zeitoper,* not when the *Zeit* favored jazz and spoofs of modern life. Three years after the academic sophistication of his *Cardillac,* Paul Hindemith set contemporary Berlin to singing in *Neues vom Tage (News of the Day,* 1929), a dandy satire by Marcellus Schiffer on divorce and media blitz; that same year brought out Schoenberg's *Von Heute auf Morgen (From Today to Tomorrow),* on Max Blonda's libretto (Blonda in fact being Gertrud Schoenberg), equally topical in subject matter but musically too extreme to give delight to many. This pair fulfilled the implications of Strauss' *Intermezzo* in strictly deromanticized forms, and if one was jolly and the other dour, Hindemith and Schoenberg were using the same means to heighten the action, specifically, the music that plays around rather than weds itself to the words . . . pinpointed notes from nowhere, wayward counterpoint, abstractions grimacing not to the cadences of the poetry but of their own accord: contraposition.

Let us seek the gist of the *geist* that informed the *zeit* and its *oper* in *Façade,* William Walton and Edith Sitwell's so-styled "entertainment" of 1923, twenty-one insouciant movements collided with twenty-one preposterous poems recited in notated rhythm. *Façade* borrows the hornpipe, the polka, ragtime, and such, forcing one back lest some spell be cast and lyrical abandon reign. How long ago *Mignon* must have seemed to *Façade's* first-night audience, which, incidentally, was not amused; scarcely had it accepted the graceful teeter-totter of "Four in the Morning," the ruminations of one Mr. Belaker, "the allegro negro cocktail shaker," than a jagtime strut intruded in "Something Lies Beyond the Scene (the encre de chine, marine, obscene)," followed right along by "Valse," with its exquisite viola groan on the offbeat like the yawn of a banshee. It was more than they could bear.

Speaking of the project years later, Edith's brother Osbert wrote, "we sought to reach a country between music and poetry, where, as on the border between waking and dreaming, new landscapes would show up with a brief but memorable vividness." Like their contemporaries disengaging the unity of word-song built up in the nineteenth century, Walton and Sitwell reached that country in dance forms and nostalgia mated with nonsense verse, a mélange not totally unlike Schoenberg's *Pierrot Lunaire,* though of course infinitely more polite.

And so, besieged by the Word, music turned on its axis in the revolution that occurs every so often, the better to reassert itself, yes, but also to service the dramatic entity as the next generation conceived it. In England, *Façade* inspired nothing but charm from Walton; in Czechoslovakia, Leoš Janáček contrived a song cycle about a peasant stripling who runs off with a gypsy girl in terms that forced the romantic *Lied* to bend to Slavic lecture yet without giving up the development of cyclic themes; in Switzerland, Igor Stravinsky set the tale of a soldier outwitted by the devil for the sardonic commentary of a seven-man chamber ensemble and three human speakers. Each composer enacted writs upon the Word, but how differently, each, and how varied is their era. *Façade* is the beau ideal of contraposition, while Janácek even in his spare last operas was never to release the vocal line from the dramatic expression we associate with traditional opera. Stravinsky, on the other hand, could force any mood on the parts of his scores, be they voices or instruments, though one cannot resist the impression that his characters aren't quite alive. How pervasive is the impression that it is the bassoon that sings, not the tenor; that the crisp scruples of the trumpet are in question, not those of the chorus; that Stravinsky in his long career never cut loose of the puppet play, of the deanimated man of modern continental European drama. When he first began composing he seemed to be taking romanticism to its logical conclusion, but the conclusions of a romantic era could not possibly be logically worked out: discovery was the romantic's drive, and there is always more to discover. Ultimately, Schoenberg's *Gurrelieder* would be the outcome, the logical conclusion, but even so it features one movement in *Sprechstimme* in defiance of romanticism's word-music compact. If *Gurrelieder* is the conclusion of the nineteenth century, it is no logical one, for Schoenberg trapped the verbal squawk of the next epoch in it, proving that romanticism's last discovery was that romanticism had ended. Still, for its reappraisal of Wagnerian music drama, *Gur-relieder* redeemed the last age; Stravinsky could only lead the next one.

But to where? The Janáček cycle, *Zápisník Z Mizelého* (*The Diary of One Who Vanished*), composed in 1917-19 to an anonymous text that Janáček spied in a newspaper, carries the Word as far as it can go in *collaboration* with music, while Stravinsky's *L'Histoire du Soldat* (*The Story of the Soldier*), which dates from 1918, attempts no collation of C. F. Ramuz' text with the music, letting them take turns, the end-rhymed parable of a soldier who trades his violin, his soul, to an evil genius in several disguises thus illustrated but never portrayed by music. Annotation, rather than characterization, is here, and parodistic ragtime and tango, too, for *L'Histoire du Soldat* is a child, or perhaps father, of its time, and more exactly *Zeitspiel* than *Zeitoper*—a theatre piece, no opera—for music drama isn't always to be "opera" anymore.

Easily the cornerstone of *Zeitoper* was Ernst Křenek's *Jonny Spielt Auf* (*Jonny Strikes Up the Band*), to his own libretto, at the height of the brawling social and economic circus in 1927. Don't bother requesting, or attempting, a definition of "jazz"— one knows it when one hears it, and in Křenek's saga of a lawless black musician from America on the loose in western Europe one heard a great deal. The rinky-dink pushbeats of the fox trot, the overwrought tango, the woodblock, brass "growl" and cymbal stinger define Křenek's fetishization of Yankee Doodle savvy, and mechanics of the popular media run rampant in the piece, as, indeed, they had just begun to in American life. Křenek marked the first appearance of his stage band for "schnelles Grammophon Tempo," and when Jonny swaggers in, saxophone in hand, his first lines are, "Oh, ma bell, nicht so schnell; gib mir eine kiss!" He spends most of the evening trying to steal a priceless violin—the better to bow the blues, as it were.

Křenek was a brilliant theatre man, and he caught the Corybantic compulsion of the epoch in his scenario, which leaps from one bizarre event to another as neatly as the syncopated bellowing of the cabaret careened from platitude to cynicism. This was Offenbach's *gai primitif* for sure, with all the bravado of the vernacular stage; "Jonny spielt auf" can also mean "Jonny cuts a swank," and that Jonny does for much of the evening. The author's best moment is the murder of the violin's rightful owner, shoved onto the railbed of a metropolitan station just as a train roars in—and before the audience can even gasp, a drop curtain plummets to the floor, cutting the business out of the story. Of what importance are rightful owners when the future of jazz is at stake? In the finale, with the violin in his grasp, Jonny materializes on a giant revolving globe of the world as the chorus, attired in red, white, and blue, tenders a straitlaced hymn of praise to almighty Jonny and almighty jazz: "The new world comes journeying over the sea in splendor and takes charge of old Europe with the dance!"

"Impudence" and "the determination to surprise" was what Clive Bell called jazz in 1921 in his essay, "Plus de Jazz," and he caught on to the secret of how jazz operates as theatrical narrative. "You shall not be gradually moved to the depths," he wrote; "you shall be given such a start as makes you jigger all over." Surely, jiggering the crowd was Křenek's intention, and the crowd did its most ecstatic jiggering back at the Munich premiere of *Jonny Spielt Auf,* in 1929 at the Gärtnerplatz Theater. Alfred Jerger, the white-glove-and-blackface Jonny of the occasion, later recalled that resistance struck when he leaped onto a piano to launch into a particularly smarmy solo, the public

silencing him with roars, whistles, and cries of pfui. Warned that
the work heralded the conquering of the world by the black race
through music, patrons had come armed with rotten eggs and
stinkbombs, and it was at this point that they bestowed them
upon the stage. The performance was not completed that night.

Kurt Weill and Bertolt Brecht were even more effective than
Křenek at rousing the public to violent commentary. In a twenty-
five-minute chamber piece, *Mahagonny,* presented at the Baden-
Baden Festival in 1927 on a bill with Hindemith's *Hin und
Zurück,* Milhaud's *L'Enlèvement d'Europe,* and Ernst Toch's
fairytale *Die Prinzessin auf der Erbse,* Weill and Brecht blasted
the innocent heel-clicking of the first three with a savage cycle
of songs on the degeneracy and pillage of modern capitalism, and
a year later they merged *Zeitoper* with timeless social drama in
Die Dreigroschenoper (*The Threepenny Opera*).

This work has since become one of the classic music dramas
of the century, no small feat for a flat and unfriendly piece that
at moments has as much thematic validity as a crackpot on a
soapbox. In place of the dance-band gleanings of Hindemith and
Křenek, Weill made the rickety jazz-combo sound an absolute
in its own right, and he further distilled the profane tone of the
score by purging any trace of romantic plush, whereas almost no
one else of his time was able to dig that deeply into staccato syncopation and still remain tonal. The easy way out was simply to
invoke the ethos of blues by flatting the seventh and the third,
especially in the melodic line; this procedure informed a whole
subclass of jazz opera, carrying one from a few bluesy moments
of Maurice Ravel's *L'Enfant et les Sortilèges* (*The Child and
the Enchantments,* 1925), a charming fantasy on a libretto by
Colette, to the Harlemite drum beating of Louis Gruenberg's
The Emperor Jones (1933), a setting of O'Neill's play that

added little more to the original script than some burdensome postverismo recitative.

Die Dreigroschenoper, however, turned a game into an aesthetic: both the music and the words dug into the anarchy implicit in contemporary world-art to birth a mode of music drama suitable for sociopolitical statement, hardnosed and simple so the workers can get it the first time. In Brecht's adaptation of John Gay and Johann Pepusch's *The Beggar's Opera,* the ballad comedy of the vulgate was dragged into the modern era, bootstraps raking the mud, any trace of its likeable commedia foundation shaved off by dialectic. The setting is the meaner streets of London. The protagonist, Macheath, is a gangster; so is the police commissioner and the leading businessman, which is doubtless why German theatres often stage the work with the entire cast in evening clothes. The focus, now more than ever, is the Word, and not only because *Die Dreigroschenoper* is basically a play into which songs have been hammered, to impress the message on the spectator. The words are acerbic, and the music is deadpan. Don't just hum us, advise the tunes. Receive the verses. (Yet can we live up to the asthetic? Anyone can hum the Ballad of Mack the Knife; who of us gets the words?)

The contrapositional technique utilized in *Die Dreigroschenoper* dates back to the ballad comedies of old (with an added steerage of social criticism, however), but it was a novelty in opera after one-hundred-twenty-some-odd years of romantic spellcasting, and turned up again, a little less the novelty, in the next Weill-Brecht opus, *Happy End* (1929), just as sour as *Die Dreigroschenoper* but not as finished artistically. Then the team tried it in a through-composed opera, an elaboration of their one-act *Songspiel* from Baden-Baden entitled *Aufstieg und Fall der Stadt Mahagonny* (*Rise and Fall of Mahagonny City,* 1930), and here it proved an unspeakable novelty, an incitement to riot in theatre after theatre, and it felt like the crossroads of opera as was and now very often would be.

The full-length *Mahagonny* presented the Marxist worldview in arioso and full-stop numbers, with parodistic classical devices and citations of the past (the refrain of the bridesmaid's chorus from *Der Freischütz*, is sung to the words, "Schöner, grüner Mond van Alabama), proper modern art for you. *Die Dreigroschenoper* postulated a sharkfest in the London underworld; *Mahagonny* placed it in an imaginary city in Brecht's private cloud cuckooland located somewhere between Mandalay and Chicago. Mahagonny, the "city of nets," is a gangster's dream town where the only crime is not to have money. This is what brings down the hero, Jimmy Mahoney. Even his girlfriend won't lend him the cash with which to save himself, and she defends her stand with a song of typical Brechtian nihilism:

> *Denn wie man sich bettet, so liegt man,*
> *Es deckt einen da keiner zu;*
> *Und wenn einer tritt, dann bin ich es,*
> *Und wird einer getreten, dann bist du's!*

> So, as one makes his bed, he lies,
> There's no one there to cover him up;
> And if someone kicks out, I'm that one,
> And if someone gets kicked, that'll be you!

For so singleminded a piece as *Mahagonny* is, it is remarkable how many shifts of mood Weill and Brecht impose on the narrative—though of course everything sardonically doubles back on itself, so that, say, the lyrical Crane Duet of Jenny and Jim manages to caress with its line, "So sind die Liebende, Liebende, Liebende," and yet slap the face with the arid pulsings of its

Bachian scoring. Similarly, the scene in which the men of Mahagonny laze around listening to music and pipe-dreaming in front of the "Here You May" Inn takes its irony from a cheap sentimental ballad played on the piano; "that is art eternal," states one of them, while Jim breaks into a nostalgic reverie on his days as a lumberjack, promptly loses his grip, and has to be restrained in a busy ensemble, the climax of which sounds like a cranky cakewalk. In *Mahagonny*, the land's end of capitalism as viewed by the Marxist, there can never be contentment, as Jim explains it, "because too much peace reigns, and too much unity, and because there's too much on which a man can depend."

Ultimately, high prices and an epic picket line end the saga of Mahagonny, and it's easy to see why the work was greeted with such disgust by so many in its day, one harried by unrest and an uncertain political future. (It may be that only Americans, with our sense of manifest insuperability and unsubvertable democracy, can shrug away rabble-rousing art.) Almost everyone was outraged by the show but the avant garde of the musical establishment, who were floored by Weill's disturbingly brilliant musical scheme. As Busoni's pupil he would of course have mastered the mystery of forms and devices, but the special personality of his sound was as original in its own way as that of Stravinsky. Both built on the past—nothing will come of nothing—but the sometimes folkish, sometimes inhuman polyphony of the Russian has no point of reference, after Weill's earliest works, with the hollow directness of the German. Weill turned the pastiche of neoclassicism into something new, and the flavors of Bach chorale, dance band, fugato trios (as in that of *Die Dreigroschenoper*'s overture), and passacaglia—even an exhumation from *The Beggar's Opera* for the modern adaptation—are melded into an individual style, a voice of the day.

The Nazis arraigned the Weill-Brecht pieces as "Bolshevik art," and in this one act the Nazis were right. Brecht's dialectic denies even a token goodness or distinction in man, and his ugly *Mahagonny* is truly meant to encapsulate the way of the free world. Such a repugnant view of life would be tiresome in repetition but for the fact that the artistry of both men is so consummate. It was inevitable that a genius of the stage and a genius of the concert hall bring the secrets of their respective trades to the practice of modern opera, and equally inevitable that their operas reflect the bitterest sort of antiromanticism. The canny Brecht would concede the fun of song to his audience but never, if he could help it, the escape; an artless pop refrain in pidgin English such as *Mahagonny*'s "Alabama Song (Oh, moon of Alabama, we now must say goodbye!)" was set by Weill for bizarre three-note patterns in the verse followed by a chugging fox trot with a busy middle voice on two-note suspensions and what sounds like bird cries in the treble. Compare this to the genuinely artless blue strains of, say, George Gershwin and B. G. DeSylva's one-act Frankie and Johnny opera of 1922, *Blue Monday,* and one comprehends the thrust of contraposition: Weill chafes and shoves, while Gershwin pets and promises.

A purer neoclassicism of Weill's age was no less outrageous but definitely out of the sociopolitical ambit. This was the quarter of the "intellectual" composers, conceivers of overall plans, musicians who acted as if they knew they were caught up in or perhaps even heading a movement. Igor Stravinsky would rank as one such chieftain, taking neoclassicism from the land of Rimsky-Korsakof (in whose style he molded his first symphony, opus one), right up to the last battle for contrapositional opera on the stage of La Fenice in Venice in 1951 with *The Rake's Progress.* Almost anything Stravinsky touched in opera turned to profaneness. His detractors might complain that he hasn't any heart, while his supporters might retort that his detractors haven't any mind.

One thing must be accepted, that the heart-on-sleeve opera of the pseudoheroic and bourgeois nineteenth century has passed on forever. Myth, mystery, satire, and loony bagatelle have replaced it, and at one time or another, one notes, Stravinsky got to all of them. *The Rake's Progress* was to stand as his only full-length opera, but he made a strong case for the one-act in such diverse guises as fairytale, commedia, farce, and Greek tragedy, and followed in the footsteps of Debussy and D'Annunzio in accepting the media mix of spoken drama, music drama, and dance drama as an expedient accommodation of the freedoms of contemporary theatre technology. *Perséphone* (1934), a "melodrama," follows the confrontation of death and life in the legend of Demeter's daughter, making her something of a Christ figure in André Gide's text, a heroine who serves her six-month term in hell not under stricture but out of compassion for its wretched inmates. Ida Rubinstein, the Sebastian of Debussy's opus, was again on hand as Perséphone, a speaking part; René Maison sang the only other solo part, and a chorus and a corps shared the obligations of tableau.

Here Stravinsky deemphasized the nattering woodwind and obtuse bassoon tangents of his ballets to redouble, in his own words, the "syllabic values" of Gide's verses. As harmonious as *L'Histoire du Soldat* is jangly, *Perséphone* disproved the canard that the new "objective" music was not expressive of anything but its own remoteness. But how can music be objective? The flute duet in *Cardillac* during the murder of the Chevalier derives its expression precisely by playing off the implicit violence of the coming act; this can only add to the suspense in what is after all something of a thriller to begin with. And as for the most "objective" of them all, atonalism, what could be more expressive of the horror and finality of murder than *Wozzeck's* famous unison tutti B's in the interlude after its murder scene?

In short, there is no unexpressive music, ever—music *is* expression—but there is music that expresses by veering away from its apparent subject, and once invented, this music had a habit of intruding in unexpected places. The Italians, for example, were unable to deal with the operas of Gian Francesco Malipiero at a time when such late romantics as Puccini and Zandonai were holding forth, for Malipiero's inventive formalism too often failed to pay off in Big Moments or scènes-à-faire. His *Sette Canzoni* (*Seven Songs*, 1920), the second third of the trilogy *L'Orfeide* (*Orpheus' Children*), is literally that, seven vignettes staged around songs that dramatize stray turns of fortune. In one scene, a mother laments her missing son and then backs away in terror when he suddenly returns; in another, a drunkard staggers around in the street chanting nonsense until an enraged husband mistakes him for his wife's lover and beats him senseless.

The tonal and dramatic anomalies of Křenek, Weill, and Stravinsky were more palpable and therefore easier to accept as fantasy; Malipiero's anomaly was that of real life, and his music, far more lyrical than theirs, nevertheless repulsed the public as being dementedly "inexpressive." But what was Malipiero driving at? Mosco Carner quotes the composer's answer: "The dramatic is what *one sees* while the music expresses that which one *does not see*." There is the source of this so-called objectivity in modern opera, and one simply must accustom oneself to it, for the artist of the twentieth century cannot avert his gaze from the chatterbox skeletons rummaging in the closets of life. No wonder Malipiero pulled such poor public relations in his youth: hard on the heels of the "easy" Wolf-Ferrari and his Goldoni farces came Malipiero's interpretation of the same repertory, *Tre Commedie Goldoniane* (*Three Goldoni Comedies*, 1926), devilishly incisive and contrapuntal detonations of the highest spirits, but the kind of music that makes assessment after one hearing somewhat premature. Nothing like the austere Hindemith or the alchemical Busoni, Malipiero debouched his classicism into open territory neither academic nor yet extrovertedly vocal; he won some appreciation, but had to wait decades for popularity.

Some neoclassicists, like Malipiero, sought to release the Word from its romantic enthrallment within the score by razing the foundation of harmony. Others took the reverse approach, designing great blocks of sound and rhythmic periods of no clear melodic outline to make statements of marmoreal bulk. Stravinsky—again, the King Charles' head of modern music, popping in at every junction point—set a Sophoclean Oedipus onstage in an "opera-oratorio" written with Jean Cocteau to be as static and granitic as possible. Premiered in concert dress in 1927 and a year later in *kothurnos*, *Oedipus Rex* revived some of the abstracted shriving of the Greek theatre, none of its principals moving so much as a step and anything of interest in its action imagined in the wings. To embalm the situation in a breathing frieze, the authors had Cocteau's text translated into Latin, allowing for a narrator in evening clothes to address the public at intervals in their own language; thus the libretto becomes a purely tonal device, a cog in a flywheel that never gets to turn.

In preparing for the first performance of *Façade*, Osbert Sitwell came up with the idea of having the reciter sit unobserved behind a screen, for fear that otherwise "personality obtrudes and engages the audience." This, perhaps, is what lay behind

Stravinsky's *Oedipus* setting, for who knows for certain if anything human hides behind those gigantic masks? Rooted to the
spot, sworn to hunt down the malefactor who has brought a
plague upon Thebes, Oedipus suffers one of the best-known personal catastrophes in Western memory, but in Stravinsky's adaptation we do not experience the Oedipus our own theatre has
taught us to look for, with the face and body of a man. And yet,
the music does not segregate us from the drama. If anything, it
has the whiplash grandeur one requires of tragedy, with the immensity of the fall conveyed in sudden diatonic simplicity. In
Stravinsky's perspective, C Major is the sinister uninvited guest
and Oedipus' realization of what he has done to himself, "Lux
facta est," finds the most terrible emptiness in a simple b minor
descent of F#-D-B-F#. The stage itself, rather than the music,
becomes the agent of contraposition in *Oedipus Rex,* for by reviving the speech and trappings of dead eras, it makes a ritual
of antiritual.

On the other hand, Darius Milhaud's setting of Aeschylus'
Orestiae (in Paul Claudel's translation), composed from 1913
to 1924, left the stage unhampered but threw his lot in with an
obstreperous polyphonic orchestra that sometimes sounds like
two different works being played simultaneously. *Agamemnon*
and *The Libation Bearers* were cut down to act-length, but *The
Furies,* as *Les Euménides,* unfolded stark-raving fatefulness sub-
stantially as Aeschylus penned it, though of course Milhaud's
vocal line was nothing like what we take to have been the chant
of the Greek festivals. Later on, Carl Orff would complete the
cycle of Western music drama in stark arioso and a largely percussive scoring that bids fair to doing what the Camerata could
not, but the classicists of the 1920s had not given up on the string
instruments or on something recognizable as a tune. Blocks of

chords battle chromatic strands of melody, all rather arbitrarily, and yet, amidst the fracas, the musical play is heard. (Later, when Orff tried his hand, one would hear nothing but the play.) Milhaud's happiest inspiration in the whole trilogy is the opening of *Les Euménides,* a slow, spoken rumination by a Delphic priestess over cymbals, castanets, tambourine, gong, and four varieties of drum.

With its conciseness and appreciation of text, neoclassicism did well by comedy, and was at its best in the mordancy of the puppet play. Yes, there was Wolf-Ferrari and his human Goldoni characters, and even Carl Nielsen and his *Maskerade* (1906), after the Goldoni of the North, Ludwig von Holberg. But these lighthearted texts brought out the heart more than the soul of comic theatre; after all, this is the era of the profane, exulting like a devil at the pretty pass man had arrived at in his history. Trapped between two world wars, with the delicate new republics of central Europe falling—or being knocked over—like china figures, the artists of the West had more than Goldoni on their minds. Writing his own libretto for the aforementioned *L'Orfeide,* Malipiero opened the triptych with what amounted to a treatise on the death of old-style comedy, *La Morte delle Maschere* (*The Death of the Masks,* 1925)—a direct refutation of Mascagni and Illica's *Le Maschere.* In Malipiero's scenario, seven tropes of the *commedia dell'arte* are retired as Orpheus brings in their replacements, seven modern maskers who, after an intermission, perform the seven songs of Part II, *Sette Canzoni.* (Part III, *Orfeo, ovvero l'Ottava Canzone,* offers the eighth song, an entertainment by puppets which, Malipiero knows only too well, is bound to bore most of the public, and an onstage public is accordingly bored by the puppets and by the end of the piece is fast asleep.)

Here's that puppet play again, pathetically faerie in Maeterlinck's hands but vividly kinetic and all for grotesquerie in the twenties. Enter, once again, Stravinsky, with *Renard* (1922), an "histoire burlesque chantée et jouée" by two separate casts, mimes onstage and voices in the pit. This fable of fox, cock, goat, and cat doing what they do in fables was Stravinsky's own concoction, words and all (translated into French by C. F. Ramuz, the author of *L'Histoire du Soldat*), and it does everything it can to enforce the glacial "touch me not" of the profane theatre. A hebetudinous march brings the cast on at the start and sends them off at the close after the usual appeal to the spectators for applause, and surprising trills and toodle prevail throughout. It's an adorable little piece, but, as with *L'Histoire*, the bite of it is big enough to sit on.

The advent of neoclassicism could not offer a final solution to the problem of what to do with romanticism, for sacred opera was by no means ready to pass on, and the public was only prepared to accept the profane in the comic arena. In Italy, to cite an extreme example, romantic opera never even weakened—rather, it rose to an apex in the postverismo era—and by and large the German operagoer was still waiting for a successor to, and disciple of, Wagner.

One solution, but by no means a final one, was offered by Arnold Schoenberg in his twelve-tone system, the ultimate in contraposition and yet, for all its formality, nothing like classicism. The man who scaled the finite reaches of romantic, cyclic composition in *Gurrelieder* saw Gustav Mahler negotiate a truce between past and present, but Schoenberg sadly forged on into the future feeling inescapably postdated. Unlike Stravinsky, who retained his roots in folk song and vertical tonality up to his final years, Schoenberg dispatched harmony altogether in a uniquely

horizontal texture, lines laid out on lines, all broken by bleeps
and sniggers and moans in wide gaps. There was no longer any
question of an opera orchestra grimacing along with the action:
the music was now fixed in a grimace from first note to . . .
well, not last. There are no last notes in Schoenberg's operas.
Suddenly, at a point neither particularly here nor there, they
have ended.

Schoenberg's first stage work was *Erwartung* (*Waiting*), to
Marie Pappenheim's highly erotic text about a woman who
searches for her lover in a forest and finally stumbles upon his
corpse. Composed in 1909, *Erwartung* had to wait till 1924 to
be heard, and even now its demands in the way of concentration
and belief are far more than the average operagoer can meet.
Perhaps the first mannerist opera to cut itself off from the pop
decoration of *Jugendstil* or legend, *Erwartung* seeks the real behind the real, the overtruth, in an apparently chaotic blather of
words and noises. No one "tune" or melodic period is articulated,
for the sum is what tells, not the parts. No tonal base, no succession, no release is offered the audience, just a continuum of
expressionistic nightmare. Critics have noted the relationship of
Kundry and Elektra to *Erwartung*'s Woman, but so decocted is
Schoenberg's operation that the relationship is that of an egg to
a chicken fricassée. Distended, harassed, this Woman and her
improbable vocal line represent the almost total disintegration
of the forms built up from the Camerata on through Gluck and
Wagner—and yet Schoenberg's atonalism is nothing but form,
for by choosing one peremptory arrangement of the twelve tones
of the scale and playing it over and over in myriad variations and
transpositions, the composer reached the last modulation of cyclic
unity.

Schoenberg's *Die Glückliche Hand* (*The Lucky Hand*), composed in 1913 but, again, not staged until 1924, pressed even deeper into the landscape of *Erwartung,* this time by moving out of the dream world into no world at all. The musician wrote his own libretto, calling for a complex pattern of colors in the lighting plot to mirror the technique of his tonal palette; as in *Erwartung* and the song-cycle *Pierrot Lunaire,* he was seeking to essentialize those tenets of modern art only partially explored in the puppetlike trances of Stravinsky's chamber pieces, in the Freudian surrender of absurdist drama, and in the Grand Guignol of such artists as James Ensor and Alfred Kubin.

Even before the turn of the century, the automatons, doppelgängers, and demon temptors of the E.T.A. Hoffman era—tokens of the unseen psyche—had been promoted from filling out other characters' stories into an aesthetic all their own, with every man a Hoffmann and every tale one of Hoffmann's fantastic tales. The Woman who waits in *Erwartung,* the Man in *Die Glückliche Hand* with the feline monster crouching on his back, even the bored couple in *Neues vom Tage* who manage to survive Schoenberg's challenges intact . . . all are subject to disturbing incongruities in the action, in their music, in the world they inhabit. However horrible the disasters in these three operas, nothing is tragic, for no tragic situation has been defined, and if the apparatus of atonalism suggests anything, it is comedy: the ironies are all paralogic inanities, and the dramatic statements cut into rather than relate to each other. The questions posed by *Erwartung*—is "she" really in a forest? Was she truly awaiting the man she discovers, dead, at the close? Is he really dead? Is he there at all?—and the answers offered by the small chorus in *Die Glückliche Hand,* murmuring dull nothings out of nowhere, are not in the catechism of tragedy.

The music in opera defines the tone, and the tone of these

Schoenberg one-acts is "comic," utilizing the diagrammatic inter-action that, in words, would amount to farce. Atonalism, devised by a self-defrocked arch-romantic, proved at last to be the most complete declaration of contraposition in opera, perfectly upend-ing the autotelic communion of musical theatre by shredding its most strategic component, the collaboration of music and words. These twelve tones and their endless progressions are the com-plement of modern satire, excluding the audience from knowing enough to sympathize, even with itself.

Are there places forbidden to comedy, the most dauntless of Thespis' protegés? Not, perhaps, in the haunts of twentieth-century Europe, when its political and artistic institutions were collapsing into an anarchy that any artist could feel. Satiric tragedy is not new under the sun, and we have the satyr plays of the Greeks to remind us of how long creators have been en-grossing themselves in the grin, as well as the gasp, that accom-panies catastrophe. But the puppets that abandoned the little theatres for the main stages of the Western world bring some-thing new with them, a corruption of the mind and a disease of tradition, and for these puppets Schoenberg wrote his operas. "Intellectual sterility," sniff the romantics; "perverted objec-tivity," sniff the classicists—but Schoenberg addresses them both. It is a different language, yet it communicates, not disjunctively but *as* disjunction. We first had to retire the pseudoheroes and their gran duos with some lady of the moment, then we had to retire the ripsnorting citizens of realism, but in arriving at mod-ern man in his modern opera the transformation had somehow taken words and music along with it, cutting them loose in order to free them. The anarchy of modern opera is not so much that of its characters and their environment as it is that of the libretto vis-à-vis the score.

Thus the "atonal" masterpiece, Alban Berg's *Wozzeck* (1925), divides its three acts into fifteen scenes, five each, each scene a set piece—Rondo, Passacaglia, March and Lullaby, Invention, Perpetuum Mobile—and each succeeding act taking up the notes with which its predecessor ceased. This, however, yields intellectual vigor rather than sterility, and neither perverseness nor objectivity compels the sheer force of drama, first-rate drama. Taking as his source a work out of the first minutes of social naturalism, Georg Büchner's play *Woyzeck*, Berg condensed the original for symmetry, but it is not the symmetry of a Busoni—one can hear this music bleed—and the highly personal quality of the singing leads one to understand that Berg had reconstructed twelve-tone technique for his own purposes: *Wozzeck* is no opera of contraposition. Yes, that grimace of expressionism is present, but on the other hand each principal appears in a distinct tonal aura that stems from Wagnerian leitmotif, and if the superb final interlude of Act III is pitched in an undisguised d minor, there are also passages of quite accessible beauty throughout; this would constitute a breach of contract in Schoenberg's one-acts.

Wozzeck: a hapless soldier who has fathered a bastard on a vacillating woman is the butt of the world's joke, experimented on by a Doctor, cuckolded by his common-law wife, and beaten by her latest flame. He stabs his wife and drowns while looking for the murder weapon. When the woman's body is recovered, a group of children halt their games to view the corpse, including Wozzeck's child on his hobby horse.

Not much happens, then, but expressionism is almost nothing but incidentals, and in the event a great deal happens; *Wozzeck* is, in fact, one of the most successful narrative operas and one of the most traditional in its musico-dramatic operation. So fragmented is much of the orchestral deployment, so overshy or intrusive the glissandos and tremolos, that the motivic groundwork is obscured at first or second hearing. But the procedure is in essence Wagnerian, as was Schoenberg's, and repeated hearings reveal characteristic themes and thematic groups—the Drum Major's little tattoo, for instance . . . yes, a military theme, with a touch of farce in its half-tone harmony, nothing more emancipated than something out of Chabrier. There is really no distinction to be drawn between how Wagner and Berg see opera other than in the enforced clarity of Berg's vocal line, its resistance to subsumption in texture, its antiharmonic precision.

Precision—that is the key to much of this new wave of the 1920s, particularly in France and Germany. What at first glance sounds like an involution of eloquence (or at any rate did at one time) is if anything mega-eloquent, a multiplicity of rhetoric and formalization that still ends up being no less instinctually "right" than the best of Verdi or Wagner. Debussy claimed music's obligation to limn the "inexpressible," and *Wozzeck* does grasp the innermost psychological corridors in Büchner more fetchingly than Berg's coevals generally were able to do with even the most plastic of librettos. The chording in *Wozzeck,* often enough, seems to ignore its tonal center or not have one at all, but the melodic pattern, interestingly, is as imitative and cyclically ongoing as anything in the *Ring.* That much of romanticism even the avant garde could not release—character, situation, idea were still being held for ransom by the ur-motive and its conduits.

Precision—that is why romantic opera found itself, at what it took to be the height of its powers, at the crossroads, and no roads led back to Rome. Precision: so the inexpressible could at last be expressed. More and more librettos will come forward that could only be read from the stage in some precise code; oddly enough, the major blast of modern precision in opera, this *Woz-zeck,* took its source from the crypt of the nineteenth century and encoded it for what amounts, very nearly, to an insane expansion of nineteenth-century aesthetics. Vocally and orchestrally, *Woz-zeck* is a fantasia in form—or, rather, a suite of fantasias—and dramaturgically, too, it tallies as a free-form developmental move-ment, skipping the exposition and having no real denouement. The first scene, curtain up on Wozzeck shaving the Captain, has really begun after *Wozzeck* has commenced, and the final scene of the children playing takes place after Wozzeck and Marie are both dead and *Wozzeck* technically over. But then *Wozzeck* isn't about Wozzeck to all that extent. Marie's more sympathetic vocal portrayal and her less expressionistic, more naturalistic dialogue render her more approachable than he—and, indeed, it is her scenes that have been swiped from the whole for concert performance.

Why, one wonders, did Berg change Büchner's ending, in which Wozzeck does not drown himself, but is seen under arrest at the morgue with almost the entire cast on stage while a police-man surveys Marie's corpse? "What a murder!" he exults. "It's been some time since we've had one like this!" And, the stage directions tell us, the protagonist stands there silently staring at the woman he murdered, "tied up, the positive atheist, tall, hag-gard, shy, easygoing, scientific." It's quite a picture, and, given the intentions of Büchner's social document, the definitive *hic jacet* for "wir arme Leut." But Berg had other ideas, and his ar-rangement of *Woyzeck* owes less to expressionism and the mod-ern drama that Büchner helped to found and more to the organi-zational technique of neoclassicism. Decades later, Bernd Alois Zimmermann found the irregularly sculpted scenario of Jacob Michael Lenz' *Die Soldaten* ideal for operatic structure even

with its tiny scenes of a speech or two—and Lenz' play was very
much a forerunner of Büchner's. But Zimmermann's elaborate
scheme allows for scenes to be played simultaneously, one or two
overlaid on others; Berg went for balance. Nothing in Büchner's
Woyzeck suggests the doctrinaire confidence of *Wozzeck's* three
acts, each of five scenes, each scene connected yet separated, as
if in arithmetical progression. But then Büchner wasn't writing
in the twentieth century.

The difficulty in grasping such a work as *Wozzeck* when it
is not done in the language of the audience is well-nigh insur-
mountable, even for those who know it well, for it is constructed
to play line by line, not—as with, say, *Rigoletto*—musical scene
by musical scene. Romantic opera, except for Wagner, late
Verdi, and a very few other exceptions, works in strophes and
aggregate contours, while modern opera is designed to adjust
to text, to recruit Wagnerian music drama on a Word-basis. Perhaps no other single event recorded in this book has proved so
strategic for the fortunes of the present day than *Wozzeck,* for
it offered a way out of *Spätromantik,* a way that led both on into
the wisdom of neoclassicism and back into the dramatic power of
romanticism.

But just as Gluck with his persistent viola figure in *Iphigénie
en Tauride* had to fight down a general resistance when he invested the drama with music, Berg had no easy time of it rein-
vesting music drama with its words. The holdouts would have
treated *Wozzeck* as they had Schoenberg's two operas the year
before, but so powerful is its impact that the piece reached
twenty-nine different theatres in its first decade. This is a sort of
comment on the remoteness of Schoenberg's methodology, for
Berg leavened the gross points of strict atonality with an elastic
sense of musical characterization. No one line in *Erwartung—*if

one line can be isolated—is heard more than once, thus mesmer-
izing the id of Western song, but *Wozzeck* is not so distantly
placed from the willing ear. A lusty country chorus and a sentimental waltz intrude on the beer garden scene in Act II; the
march of the soldiers in Act I shows its blood type with Křenek
and Weill, and there is even a hint of Humperdinck in the lullaby that immediately follows it.

This is not to suggest that atonalism works best when it least
offends, but that it does lose its efficacy when it disposes of all
mood and action. Schoenberg was to pull back a bit from his
dead end and create a work of eminence in *Moses und Aron,*
but it fell to his pupil Berg to abstract the atmosphere of atonal
contraption within the still viable mainstream of expressive—
romantic, if one will—opera. Syndetic Berg: he rebuilt bridges.
Both in *Wozzeck* and in the later, more specifically dodecaphonic
Lulu, Berg launched a sortie into the satiric spotlight to haul it
back onto the traditional stage.

Well and good. But from now on, things cannot stand as they
stood. The Word, reconnoitered first by Debussy and Janáček
and now, again, by the jazzmen, neoclassicists, and atonalists,
will never cede to the sinfonia and canto. The bimodal plane of
satiric tragedy has revealed itself, if not in *Wozzeck,* then certainly in *L'Histoire du Soldat, L'Orfeide,* and their children.
Come back, Wagner: nothing is forgiven.

V
The Idea of a National Opera

Words of convenience, "national opera." The term is as handy as "among those present" for lumping together the odd ethnic groups that don't get chapters of their own. However, it isn't only the Russians, Finns, and Herzegovinians who rate as nationalistic; much of their patriotic technology conduces to the simple default that international singers aren't expected to master Russian, Finnish, or Serbian, and thus the operas composed in those tongues remain remote, exclusive, and, it is assumed, ethnicer than thou.

Yet reflect. Have we greater evidence of an operatic repertory *pro patria* than that of, for example, Germany, with its *Der Freischütz* and *Die Meistersinger*, its *Feuersnot*, its *Hänsel und Gretel*, its *Mathis der Maler*? What will you have—national subject matter, national poetry, national song, national artistic ethos, or even just national ism? All are there in this distinctly non-Slavic, non-Finno-Ugric sphere, and in addition there is the possibly even more nationalistic English opera. There is more to this question of national opera than the linguistic inflection of the vocal line such as we note in Dargomizhsky and Mussorksky . . . and, for that matter, in Lully and Gluck as well, in the days when the French spoke of a French national opera.

In other words, every country has its national opera, though the operatic output of some countries has proved "national" in different ways from those of others. For musico-linguistic contour one may journey to Russia and for musical settings of national poetry to England, while Italy takes stage for its national patriotic crise (in Verdi, especially, no matter what the locale), and Spain in its tiny operatic history wins the palm for an attempted assimilation of national folk forms. But hold. What people knows more patriotic glory than the English? Who adapts folk style more frequently than the Russians? And who made more of a riot about national cadence than the French? It all crosses over, and there are no rules. Let us then accept that the more exotic languages of Europe have no stronger suit in the matter of national opera than those of the more accessible Latin and Germanic peoples, however isolatedly national they may be in both frontier and spirit.

National opera, as it happens, is very healthy, nationalism itself being a lot more productive than most isms, more virile than vegetarianism, nicer than sadism, and righter than Marxism. Every artist in the nation can join—everyone must, in fact, or the national art would be reduced to riding like a beggar on other nations' wishes. The previous chapters have tended to numerous examples of national opera . . . in the legends of Rimsky-Korsakof, devoted in *Sadko* and *The Legend of the Invisible City Kityezh* to the ongoing aspirations of archetypal belief, or in the historical plunge of the D'Annunzio generation of Italian composers, rooting through the past for views of the insurgent patria, or in Pfitzner's mystical treatise on the continuity of German music in *Palestrina,* so very German in attitude despite its Italian derivation. What is most invigorating, however, is to learn how each nationality of musicians and librettists deals with each Continental development in art, for by the mid-1900s, the accumulated trial and error appears to arrive at some few plateaus in the operatic aesthetic, some few definitions—and the elements of the manufacture may be perceived, discrete, in their places of national origin. Let us tarry in this arena, briefly, and then, just to be perverse, let us do what is always done and lump the ethnic groups together after all.

First, we must address ourselves to those whose sense of national opera starts and ends with rustic charades involving romance, high spirits, a dash of folk wisdom, a quotation or two from the public domain, and a brigade of peasants, preferably Slavic—something, no doubt, by Bedřich Smetana. But this is less national opera—which aims at a celebration of cultural ambitions—than folk opera, which leans to a simpler celebration of the cultural roots. National opera moves centrifugally, radiating outward to a free field of operations, while folk opera moves centripetally, inward to the core. Naturally, the folk don't write operas themselves (nor do they get to many as spectators), but their context informs folk opera, especially in the use of vernacular devices and a tang of cultural—seldom individual—destiny. *The Legend of the Invisible City Kityezh,* with its symbolistic principals and Christian pantheism, is folk opera; *Sadko,* built around a self-made hero, is national opera. The one proffers the inevitability of a people; the other, the triumph of a man.

The folk are one thing and the artists quite another. As it is the artists who write the folk operas, technically there is no real folk opera as such, only opera conceived to echo the folk through the priorities of a particular composer and librettist. But the national aggregation of artists through the years does create a folk tradition, if of a very elitist sort, and as era sinks or surges into era and forms develop, this tradition becomes the national mad money, to be drawn on as needed, and, also needed, to be invested for further profit. And as Western nations built up their balance, they inadvertently set up a pool, also to be drawn on, and thus has opera grown, first regionally in vertical rays, and then ecumenically in a horizontal network. Such is opera, knowing at last no one nation, though once it was strictly *tragédie lyrique* and *comédie mêlée d'ariettes*, or *opera seria* and *buffa*.

We come to this because, although the borders were pulled down long ago, form is still the first consideration in art, and now that the miscellany of modernism has failed to make any distinct account of itself, it may be announced that the form known as opera has reached its ultimate completion and can do anything it might have to in order to justify the ways of Godot to man. In *The Rake's Progress*, it can demonstrate that profligacy will be punished because the debatable moral code is proved by the undebatable code of classical figuration. In *King Priam*, it declassifies one's prescribed horror at the inexorability of fate, showing it to be meaningless by building up the character of the king to heroic stature and then, when he is killed at the end, dismissing the supposed tragedy in a few measures of note rows that don't even amount to a shrug. In *The Bassarids*, it finds the human loophole in hieratic myth with a loony pastoral turn, "The Judgement of Calliope," enacted by the Thebans who moments before were assaulting the very order of life with Dionysian incursion. In *Die Soldaten*, a junction of interrelated events in separate, simultaneous actions challenges the temporal and spatial limitations of theatre with such conviction that intellectual stimulation is forced on the spectator even as he yields to passive sensation, and art, for once, is space as well as containment.

Form is everything. What if *Die Soldaten*'s onslaught of images had been put to the deliberately insightless progress of the rake, or if the rake's moral epilogue came prancing in on the murder of Priam? It is not just tone at stake here, for mood is interchangeable according to the music, but dramatic pattern is exact. Back in the days when one had to write "revolutionary" operas to get anything done, form was less strategic because it simply *was*, but since then we have seen it share with the words and music in shaping the statement of art—form which is no longer the specie of the realm and the era but a thing which is chosen. Form is where the battle of the sacred and profane was always headed . . . form of one, of the other, or of both at once.

Form, too, is an expression of national sensibility. The light comic intermezzo, for example—the one grounded in reality, farced in and farced out, so to speak—is purely Italian no matter who writes it. But employ this form in altered tone, in fantasy or grotesquerie, say . . . and it is Italian no longer, off its poise and out of sort. Arnold Schoenberg's *Erwartung* is, one might say, an intermezzo; but the true intermezzo provides a break in the tragedy, a lift from the depths dropped between the acts of grandees in language fit for all to comprehend. *Erwartung* is its own "tragedy," told in elitist phrases, a grotesque rather than a comedy—though worked out in the *rhythm* of comedy—and, in fine, no intermezzo: it stands between nothing, for it already is. The intermezzo thrives in a poetic age, as the complement to the poetry, but ours is a dramatic rather than a poetic age, a time of grotesques, and the intermezzo's comic component has been subsumed in the whole idea.

That the intermezzo grew out of grotesque comedy—the masks—is not a paradox, for timechange is translation. But the spirit of contiguous clarity endures, so that a work such as Ermanno Wolf-Ferrari's *Il Segreto di Susanna* revives not the masks but their original meticulous whimsey, and *Susanna*, too, is national opera in its uncompromised form, in its sense of roundness, of balance, of consummation. In this spirited one-act (the husband suspects that his wife's secret is a lover, and he

eventually discovers that the mysterious vice is in fact the cigarette), no question is raised that cannot be answered, neither in
the tiny overture, nor in the husband's accusing queries, capped
by clinching pizzicatos or a woodwind button, so!—and note the
precision of Enrico Golisciani's text, here at a tender moment in
the emprise:

> GIL: *Là nel giardino* . . .
> SUSANNA: *Pieno di sole* . . .
> GIL: *Molti sospiri* . . .
> SUSANNA: *Poche parole* . . .
>
> There, in the garden . . .
> Radiant with sunlight . . .
> So much to sigh over . . .
> Little to say . . .

Is it life in diatonic miniature? No, it is a miniature set into life,
with a touch of Arabian chromaticism in the smoking of the forbidden cigarette, modernism in both action and music.

It has yet to be proved conclusively, but may at least be considered, that the freer societies enthuse less over national/folk
opera than do the static societies. Certainly Switzerland, which
is one of the freer countries, is weak in its *Staatoper*. On the
other hand, England has a patriotic repertory that rivals any, including that of Russia. One thinks, immediately, of that wonderful moment in Benjamin Britten's *Billy Budd* when the Sailing
Master and the First Lieutenant sound off to Captain Vere on
their low opinion of the French, "those damned Mounseers!"
"England for me," they sing—"British brawn and beef"—in music that pipes one flush against the spine of John Bull; Burke
and the younger Pitt speak up with them. Even when Stephen
Oliver's *Tom Jones* (1976) commences its Fielding with a pro-

logue of Olympian deities straight out of seventeenth-century Venice, this is still by way of national tradition, going back to the roots of English opera in Italian geniture.

Yes, England. How about Ralph Vaughan Williams' *Sir John in Love* (1929), a swaggering, lyrical bash based on *The Merry Wives of Windsor?* The composer wrote his own libretto, and with Shakespeare to draw on as well as bits from Jonson and Campion to scatter about the plot at interpolatedly musical moments, there is a rich and reinless poetry to sing such as one seldom sings in opera. *Sir John in Love* is the literary sort of national opera, culling its gems from the national library (plus some haet's worth of folk tunes), art to art, as it were—for literature, in case one was wondering, does have a place in opera, whether adapted from a source or true poetry of itself, born for the musical occasion. For *Sir John in Love,* Vaughan Williams had Shakespeare's *Falstaff* to guide him, but it is all England's Falstaff, too, now, and even while celebrating the national appetite, Vaughan Williams makes it his very own Falstaff, though he is not without his rivals and, in the case of Verdi and Boito, his superiors.

One may be hard put to define "the English voice," but it is instantly recognizable—witness the similarity between Vaughan Williams and Britten in recitative—and as for *Sir John in Love,* no one but an Englishman would have written the joyous dance measure that opens the work; or the jibing "When Daisies pied" (From *Love's Labour's Lost:* "Cuckoo, cuckoo—o, word of fear, Unpleasing to a married ear"); or the ensemble "Beauty clear and fair" (reminiscent of Vaughan Williams' *Serenade to Music),* which opens up into a waltz tune with a hint of Messager in it—but then French and English music do have a few notes in common. The most striking thing about *Sir John in Love* is its wealth of melody, marching out next after each like a national parade. No masterpiece of dramaturgical thrust, the piece is virtually blissed-out on tunes, and one wonders, with such vivacity and invention as the tradition of English song reveals, why English opera took so long to get going, especially given the vitality of English theatre.

Sometimes foreigners usurp the literature destined for national opera houses, so when W. H. Auden and Chester Kallman made their operatic adaptation of *Love's Labour's Lost,* the text was handed over to Nicholas Nabokov, and after an English-language premiere in 1973 in Brussels by the visiting Deutsche Oper, was quickly translated into German for use at the company's home base in Berlin. This is, surely, an unwieldy cosmopolitanism, for whether or not *Love's Labour's Lost* offers truly operatic potential, it is in its English poetry that the work must tempt a composer. Auden and Kallman excised much business and the characters who busy themselves with it as well as Maria and Longaville, and made Moth over into a sort of trouser-role Puck, all of which is writ on the license of adaptation. But Nabokov inclines rather to evocative pastiche, forcing a delicate Elizabethan romance into the twentieth century by crook rather than hook. Still, as long as one is doing the literary school of opera, one might as well do it as well as the authors did the strange final scene of Shakespeare's original, when the whoop and clatter of comedy is overturned for elegy; in these last moments, the composer rose to meet his material, pastiche forgotten, and after the lords and ladies have separated, for "Jack hath not Jill" for a twelvemonth, Moth closes the piece with "When icicles hang by the wall." Then the music ceases and Don Armado delivers Shakespeare's final line, speaking it. How much more suggestive it is in an opera than in a play: "The words of Mercury are harsh after the songs of Apollo. You that way; we this way."

It is one of opera's great qualities that it can scrounge its substance from landmarks of fiction and theatre, playing upon the audience's previous knowledge of said works and illuminating them from the vantage of musical counterpoint—but the landmarks and a popular knowledge of them is requisite first. The Italians more or less devised opera and the French contributed the focus of its intellectual incubation, yet both peoples lag behind most others in the self-conscious dialogue that connects the musical stage to the national literary heritage. For example, in the interwar years when modern nationalism was peaking, it was to be expected that composers would seek out the bounty of their native writers to develop musically. Prokofyef adapted Fyodor Dostoyefsky's *The Gambler;* Janáček, Karel Čapek's *The Makropulos Business;* Leevi Madetoja, a play by Järviluoma that was transformed into a benchmark of Finnish opera, *Pohjalaisa* (*The East Bothnians*); de Falla, an episode of *Don Quixote;* Amadeo Vives, Lope de Vega's *The Discreet Lover;* and Alban Berg, Büchner's *Woyzeck.* These composers aimed at renewing their literary benefactors and themselves with adaptation, and for the most part they did. But in Italy and France at the time, subjects for operatic treatment were more likely to be drawn from foreign traditions than from those of home. Zandonai went to Lagerlöf, Flaubert, and Alarcón, Respighi to German and Danish playwrights, one Alberto Ghislanzoni surfaced with a *Re Lear,* and Alfano set Tolstoy, Indian legend, Balzac, and Rostand's *Cyrano de Bergerac* (in French, no less). Even when burrowing into Gozzi for *Turandot,* Puccini footled about with pentatonic platitude and lost much of the flavor of the original's fantastical comedy while emphasizing the simply fantastical. As for France, the classical revival drove such as Milhaud, Satie, and Roussel not into Molière or Corneille, but to Euripides, Plato, and Sophocles—although one wickedly parochial outing just after the war,

André Bloch's *Guignol* (*Punch and Judy*, 1949), resuscitated the insolent musical comedies of the Parisian fairgrounds with dance, pantomime, gaucherie, and grimaces and, neoclassically enough, miscellaneous homage to Debussy, Weill, and Stravinsky.

Actually, the Italians did undergo a patriotic scare in the 1920s, on their way to fascism, and whereas the Mascagni-Puccini generation had sworn no fealty to Italian subject matter, their successors Pizzetti and Montemezzi were at least open to Italian antiquity for their romantic campaign. Of all his oeuvre, Puccini's one go at a national piece was the one-act *Gianni Schicchi,* with its intimations of the ascendance of the meritocratic Renaissance, for his two other "Italian" operas, *Tosca* and *Suor Angelica,* have no more Italian rhetoric than any given Italian opera naturally would: they exhibit canto, but no particular patria. Montemezzi's *La Nave* showed a keen feeling for historical processes obtaining in preducal Venice, the man-to-man code of the tribe grasping the propect of statism—and this in D'Annunzio's fanatical verse—but *La Nave* was exceptional, and Italians displayed more interest in buying best-sellers than national heritage as stock for the lyric stage.

Simpler in concept than the literary opera such as *Sir John in Love* and *Love's Labour's Lost* and the historic opera such as *La Nave* is folk opera, the grassroots school of music drama. Here, not surprisingly, we encounter the folk in general, very much in general—its argot, its morality, its inevitability. As mentioned earlier, folk operas tend to pull back when confronted by individuals; this genre prefers the continuity of anonymous regionals, and where there are individuals, leaders emerge and continuity is broken, anonymity shaken. Folk opera prefers types, or even the petits bourgeois of soap-opera pseudo-tragedy (real tragedy, implying the existence of heroes, is unsuitable).

It is necessary to understand the distinction between the national and folk opera constructs in these pages because the implicit "common man" slant of folk opera has advertised it as useful propaganda in this century of common men, even as national opera has come more and more to concentrate on individuals. The thrust of contemporary opera has come, as we shall be seeing, to a determined strategy of heroic resistance. Where the Greeks predicated destiny and character weakness as the hero's nemesis, the would-be Greek opera of the Camerata reduced destiny to the happy inventions of a deus-ex-machina, leaving no room—no need—for heroes. But since the intrusion of romanticism, the asymmetry of the world has been revealed along with asymmetrical protagonists, strivers doomed to see only contradictions, to be failed not by themselves so much as by the formless System. This is why formalism is so necessary to modern composers—*something* there must be that keeps its code, that orders its boundaries and patrols its community. Modern opera predicates the disordered world as the destructive force. Exterior, rather than interior, forces will bring down today's singing hero, and tragic resolution becomes less easy to define. Again and again, modern "tragedy" will stem from a dysfunction of identity in a paralogical world of unheroes, instead of from a coherent but flawed identity in a world that, but for heroes, would work.

Identity is never in question in folk opera, however. Using the strophic naturalness of the folkish strain and the imagined wisdom of the natural folk, folk opera serves as a plank in the platform of the statist demagogue—*Zeitoper* and *Realpolitik* at once, one thing to all men. But on the other hand, folk opera can be anything it wants to be, all things to other men, and it stands paramount in the priorities of composers and librettists in the outspokenly antistatist U.S., no less now than formerly. A classic early sortie in the folk opera campaign was Scott Joplin's *Tree-*

monisha, a startlingly unimpressive work recommending the emancipation of American blacks from superstition through literacy to golden rule morality. Published in 1911 and auditioned in Harlem in 1915 but not staged until 1972, *Treemonisha* has been hailed by many for a presumably infectious use of the story ballad ("The Sacred Tree," for example) and ragtime (the monomaniacal "We're Goin' Around"), but in truth the reasons for its recent success in America are better known to the promulgators of the ragtime revival than to Apollo Musagetes or Thespis. Only the penetration of expression, line, and form earns a tally, and any number of comparably "primitive" musical works show a smarter heft of musico-dramatic communication, from Jerome Kern's *Show Boat* (1927) to Stephen Sondheim's *Pacific Overtures* (1976).

The true linchpin in American folk opera is George Gershwin's *Porgy and Bess* (1935), which was one of the first through-composed music dramas of international class to be presented under commercial auspices on Broadway, rather than in the revolving or stagione repertory of an opera company. A Broadway revival in 1942 tried replacing Gershwin's recitative with spoken dialogue, but the opera has almost invariably been heard as written (if cut for length), and New York's idiot critics cheered its revival in 1976 as some sort of breakthrough forty years after that breakthrough had been accomplished—just as idiot music critics in 1935 jeered at its naiveté, not realizing that unsophistication is the goal of the form. Anyway, Gershwin's score and orchestration are far from naive in construction, not least because in catching the plaint of black vocalism he endowed that old bugbear, recitative, with new vigor and flexibility decades before America could even be said to have much operatic creativity.

Folk operas usually boa,c original storylines or stories leased from tradition; *Porgy and Bess,* however, derives from the play that DuBose Heyward and his wife Dorothy made of Heyward's novel, *Porgy.* But the exception proves the rule, for Heyward was something of a folklorist in black culture, and from novel to play to opera, the sense of racial continuity, of neither rising nor falling but surviving in isolation, is omnipresent. There is nothing symbolic about the piece. *Porgy and Bess,* the baritone and soprano principals (in two of opera's most voice-mauling parts, incidentally), haven't the anthropomorphic luster of the two love-haters in *La Nave,* who register the clash of contrary psychologies (for which read contrary historical forces) in their relationship. Yet, at *Porgy and Bess'* finale, when Porgy decides to take his goat-cart from Charleston to New York to find his departed woman, a feeling of outreach and stability overwhelms one—yes, even in this hopeless quest, for the spell cast by Gershwin's music makes it, for the audience, anything but hopeless. The feel of black song—the spiritual and the "shouting" gettogethers—pervades both novel and play, and the subject shouted for opera, but on its own terms, idiomatically. The circumstances of the plot may be bizarre, but the strength of Porgy's farewell, "I'm On My Way," focuses belief on his permanence and that of all his race. One can't be sure that he will ever get to New York, but the music is so exhilarating that one accepts. The folk endures.

One of folk opera's charms is its cultivation of the vernacular voice, especially in melodic terms. In the hands of someone as brilliant as Leoš Janáček, tiny folk themes are woven like chatter into a symphonic web and endlessly developed in the orchestra rather than simply quoted for local color, though there is something to be said for the simplistic quotations, too. After all, when an opera spends much of its evening restating, in little, the tenets of centuries' worth of popular musical happenstance, even so gullibly as in *Treemonisha,* one begins to feel that art is less artificial than preternatural and, like Pontius Pilate, is impelled to ask what, or who, truth is.

The early work of Carl Orff is a case in point. Without literally citing the tunes of old (though all of *Carmina Burana* is rediscovered medieval text), Orff nevertheless experimented with heavy-footed bauernlieder, ultradiatonic and tautologous. His two adaptations (libretto as well as music) from the brothers Grimm, *Der Mond* (*The Moon,* 1939) and *Die Kluge* (*The Wise Woman,* 1943), connect modern opera back to the miracle and morality plays of the Middle Ages, when everything was clear—and when the population, except for a minority of nobles and clerics, was exclusively folk. The folk are harder to find nowadays, and perhaps Orff was making a point in seeking them out so openly in these fairytale operas, so unlike the chromatic, passionate *Märchenopern* of the Humperdinck era. Eventually Orff was to reroute his energies in almost unlistenable settings of Hölderlin's translations of Greek tragedy, thus preparing elitist art about (and presumably for) the elite. But in *Der Mond* and *Die Kluge* he hymns the folk in folkish terminology.

Die Kluge, "The Story of the King and the Wise Woman," is the simpler of the two, a straightforward moral lesson in which greed in search of wisdom learns that true wisdom is love and greed is nothing. The monotonous repetitions that Orff introduced in *Carmina Burana* show up again here, most effectively in the refrain of the Wise Woman's lullaby, "Schu schuhu, schuschuhu," and much interest is provided by three emissaries of the profane, vagabonds who enjoy a field day of biting intermezzi, and who at one point bellow a drinking song about the breakdown of virtue that is in its sound about as ugly as German life must have been at the time, 1943. What with the obtrusive vagabonds, the insistent, end-stopped tunes, and the Brechtian production invariably afforded it, *Die Kluge* offers a slim but vivid excursion into antiromantic romance, a tidy package.

Der Mond, however, takes up more room than its successor, even in its seventy minutes of playing time, for *Der Mond* is subtitled "a little theatre of the world," and as such means to fulfill the implications of folk opera to the utmost. *Die Kluge* is kitchenette Grimm with its tiny cast and amiable plot; *Der Mond* is the whole world—the quick, the dead, and the immortal. Disguised as one of those "how the fox got his tail" fables, and given some epic distance by the use of a narrator, Orff's opera passes over into the nonsatiric grotesque of the phantasmal, as universal and wide ranging as any five-act festival piece.

When *Der Mond* begins, in the memory of human untime, four peasants from a country where nothing lights the night travel to a land where the moon, an oil lamp, hangs in a tree. Stealing it, they take it home, where they exact a fee from their neighbors for maintaining it, but as each dies, he cannot bear to part with his share, and quarters of the moon are interred with each body, the light waning until the last quarter is buried with the last peasant and the land is again black at night. Then, in the world of the dead, the four awake, relight their moon, and all the dead awaken with them to take up the occupations that passed their lifetimes—gaming, drinking, fighting, and sinning. Man is good for nothing, and nothing ever changes. With no more on their minds than creating a ruckus, the dead attract the attention of Peter, an old man who keeps heaven in order. Peter joins the party, organizing a marathon drinking bout until the dead pass out, dead again, and he carries the moon up into the sky to hang it where it belongs. And that's how the fox got his tail.

The simple story is charged by Orff with the responsibility of the sacred, drawing the audience into it in faith as it teaches. Rather than build the whole structure on folk verses, he gives diatonic strains to the peasants but Mahlerian complications to the world of natural forces—time-passage, night, death, and life. Less storybook opera than metaphysical exhibit, *Der Mond's* little theatre of the world plays fantasy in realistic terms, giving the peasants lusty Bavarian idioms to mouth as the world's wheel takes a mythic turn. Thus has contemporaneity altered the function of the bartered brides, retrieving destiny from the dreams of the romantics but applying it to the aggressive naturalism of the low stage.

Destiny and the signature of the vernacular fund modern folk opera, even when couched in the simplistic terms of the folk hero. Davy Crockett, for example, has turned up in a number of American operas, most of them introduced in the amateur precincts of a university theatre and quickly forgotten. However, Jerome Moross' one-act, *The Eccentricities of Davy Crockett*, satisfies the demands of adult technique while retaining the oafish freshness of its source, the country narrative. The last third of a triple bill of largely choral, choreographed operas entitled *Ballet Ballads* (1948), *Davy Crockett* follows the legend from frontier birth through courtship, marriage, war, a flirtation with a mermaid, the conquering of Halley's Comet, and a populist ticket to Congress, to the Alamo massacre and apotheosis. The charming piece overflows with melody beholden to yet not limited by the vernacular, thus drawing on the national treasury and contributing to it at once.

All three of Moross' *Ballet Ballads* benefited from the sharply worded librettos of John Latouche, especially the bizarre second entry, *Willie the Weeper*, a look at the dreams that free the dope addict from "the chiselers and the chasms and the store-bought spasms" of city life. *The Eccentricities of Davy Crocket* calls for modal ballads and the figures of the mountain fiddle, *Willie the Weeper* for keyboard boogie, and the first part, a revivalist account of *Susanna and the Elders,* for the dramatic balladeering of the camp meeting; all three are united by a tonal relationship, specifically, the flatted third that can be traced as an ur-strategy of American music (and that Křenek used so prominently in *Jonny Spielt Auf* to divine the ethos of jazz). Each of the trio has its own motto theme employing some configuration of notes that expresses, first, the major third, and then, immediately, the minor; it's about as American as one can get. (An example for the neophyte: recall the first line refrain to George Gershwin's song, "Somebody Loves Me": the major triad is stated on "I wonder," and the blue, flatted third follows on "who?.")

Subtlest of all in the national/folk pantheon is the musician who uses the national cadence not for itself but as a means to the end of artistic transformation, a method pointed to in the early days of the Camerata- and Académie-inspired music theatre, and later developed by Dargomizhsky, Mussorksky, and Debussy. Now comes the next level, that of Leoš Janáček, who specialized in symphonic developments of themes that derive from spoken Czech, from the rise and fall of the language. His scenes, his acts—his operas, even—take off from tiny melodies that quickly undergo a myriad of metamorphoses, and as Janáček advanced in his technique his operas turned more conversational than operatic, with both words and music combined in a sort of polyphonic text, the orchestra, as it were, talking back to the singers in their own tongue.

A tonal complement to the endless inversions and transpositions of the twelve-tone method, Janáček's dialogue of voice and orchestra has its roots—as what doesn't?—in the unity of the Wagnerian theatre, not in the staging unities but in those of composition, music and text sown in the same furrow. For example, the second act of Janáček's *Jenůfa* opens, after a thud of C sharp octaves in strings and tympani, with a theme played by the bassoons in thirds that careers anxiously back and forth on adjacent notes between c sharp minor and b minor. After a few hushed statements, this is turned, fortissimo, into a lightning cry of c sharp minor and b minor chords, capped by a rising seventh built of two fourths—and it is this seventh that dictates the profile of much of the second act's material, both in the voice and the scoring. It is articulated again in the first passage of recitative, first by the clarinet, then by Jenůfa's mother, the Kostelnička, and, after being passed from instrument to instrument, it creeps back in in the oboe at a tender key change to B Flat Major even as the vocal line has taken up other figures. From anxiety and anger it changes along with the Kostelnička's state of mind to regret: "And I was always so proud of you," she tells her unmarried but pregnant daughter . . . and her phrase, sure enough, arches upward from g sharp to g sharp only after pausing on the seventh, f sharp. Further transmutations enhance a movement, now in E Major, that grows in tranquility, until at last, when Jenůfa says goodnight and makes her exit, the bitter seventh of the prelude has resolved into a heavenly melody in F Sharp Major.

Literary, historic, linguistic, folkloric . . . what is it about opera, all told, that is necessarily individualistic or national, and need we mark it at all? Does it matter whether opera's "heroes" are protagonistic or pluralistic? Do we care what a work's intentions are in regard to tradition? For all the composer-librettist partnerships there have been, for all the ancient emphasis on singers' star quality and recent emphasis on the recreative faculties of stage directors and conductors—for all the *collaboration* of the living form—we still speak of *Verdi's* operas, of *Massenet's* operas, of *Henze's* operas. We know that Verdi regarded himself as an Italian democrat, that Massenet preferred a promising theatricality to social statement, that Henze's leftist politics have come to dominate his aesthetic to the detriment of his art. We know, in other words, that artists are attracted to one kind of "hero" or another: if it matters to them it should matter to us.

We raise the question here, in the middle of the chronicle, to achieve some little perspective in the matter of content now that neoclassicism has arrived and a taut freedom of option is in the air. The rapport of form and content once imposed limitations on opera; now that form is about to be anything, content, too, is an open field of possibilities, and, for example, what Smetana and Rimsky-Korsakof were bound to do with their folk operas in the nineteenth century is far removed from what like-minded composers are free to do today. Let us note the peculiarities of national opera, nation to nation—no!, man to man—as the years pass and the pages turn, for the question is of some interest.

It is time to examine the so-called national operas proper, and there is no better place to begin than with the Moravian Janáček, for his six major operas have only recently surfaced on the international circuit as forming one of the few outstanding testaments of operatic consummation in the century—and surely the real work of a book such as this is to reveal the "other" Puccinis and Strausses. As it happens, Janáček was far more the national composer than either Puccini or Strauss, for his operas present a range of Czech and Slovak display. *Jenůfa* offers a study in contemporary village life, *The Excursions of Mister Brouček* dips into fifteenth-century Prague at the time of the historic Hussite victory over the Catholic army of King Sigismund (even unto authentic fifteenth-century Czech and a Hussite anthem), and *The Makropulos Business* is a setting of a play by the foremost Czech playwright of the day, Karel Čapek. Even *From the House of the Dead* refers to national dialectic, for its Dostoyefskyan venture into the prison system of Siberia reflects the fear of Russian totalitarianism that was so much a part of the life of the peoples who lived between the Sudeten Mountains and the Danube River.

Jenůfa, Janáček's first success, is more exactly *Její Pastorkyňa* (*Her Step-Daughter*, 1904), to the composer's own libretto from Gabriela Preissovà's novel, and his most accessible work, of an instantaneous melodic appeal and, for all its compositional subtlety, almost conventionally "operatic" in its verismo horrors. Yet in *Výlety Páně Broučkovy* (*The Excursions of Mister Broucek*, 1920), Janáček's next stage work but one after *Jenůfa*, he was already experimenting with the bizarre concoctions of modern fantasy and the attendant disorientations of modern music. In *Broucek*, no sooner is one tune launched than another intrudes in polyphonic aggression, and recitative beats harmless melodies to death. Then, too, while *Jenůfa* is a heartfelt song of love and crime such as Puccini so admired as raw material, Mr. Broucek's two excursions take him into satiric fantasy to expose the barren, cowardly stodginess of the city burgher.

Drunk on beer one night, Broucek (a diminutive for "beatle") sails off to the moon, where he encounters a colony of artsy-for-art's-sake bohemians who are horrified by his earthiness and whom he cannot understand, especially when they register horror at his determination to eat a sausage—cannibalism, they term it. But if unable to fathom art, even of the poetastical lunar variety, Broucek is equally out of his depth at the heroic hour, for after an intermission another bout with the bottle sends Broucek back in time to the eleventh hour of one of Prague's decisive trials, the Catholic-Hussite War. Full of fake swagger, Broucek catalogues his gallantry after the smoke has cleared, but his lies are exposed and his countrymen throw him into a barrel for burning. Luckily, Broucek wakes up whole (still in his barrel, albeit) and in his own time, having learned nothing from his two excursions.

"It was strange: when someone spoke to me, I might not have understood his words well—but the melody of the speech! That was it. I immediately knew what was in him: I knew how he was feeling, whether he was lying or upset, and when he spoke to me—it was a conventional discussion—I could feel, I could even hear, that the person was probably crying inside . . . the magic of the human voice!" So Jan Matejček quotes Janáček, and *in hoc signo* Janáček conquered, through the logic of the Word, through *logos*. How right it was that Janáček's great splurge of creativity came when he was into old age, for not till then had the second coming of the Word made timely the inventions of a Janáček. His next opera, *Káta Kabanová* (1921), was another domestic-contemporary piece, this one based on Alyexandr Ostrofsky's play, *The Storm*, but by then the composer's sense of "the melody of the speech" informed every aspect of his craft, more so than in *Jenůfa*, so that even purely orchestral moments are part of the linguistic-musical contiguity and the act preludes seem to speak in Czech.

Having suffered the tribulations of a helpless musician waiting on the efforts of floundering librettists during the writing of *The Excursions of Mister Broucek*—fully five men labored on the first half, the moon adventure, alone—Janáček compounded his own scenario out of Ostrofsky for *Káta Kabanová*, giving free play to his strong sense of the interaction between human psychology and natural forces that founds or at least decorates all his narrative art. *Jenůfa's* second act climaxes with the fortuitous slamming open of a window by a gust of wind—"as if death were looking in!," as the Kostelnička puts it—and *The Adventures of the Vixen Sharp-Ears* deploys episodes illustrative of the cycle of nature; in *Kát'a Kabanová*, the locale, a Russian village, is dominated by the Volga River, ceaselessly flowing through the pages of the score, and the plot hinge is a storm that drives the unhappy Kát'a to confess her adulterous love to her husband. Further, it is the Volga that provides the final tragedy: rather than survive the loss of her banished lover in the parochialism of small-town life, Kát'a throws herself into the river. As the curtain falls, the hateful mother-in-law who has precipitated the heroine's collapse is seen punctiliously thanking the good people who helped retrieve the corpse, bowing to them, thanking them, a rock of dignity and protocol.

Příhody Lišky Bystroušky (*The Adventures of the Vixen Sharp-Ears*, 1924), known in English as *The Cunning Little Vixen*, is in a way the most essential of all Janáček's operas. With characters drawn half from the animal kingdom and half from human archetypes of country life, *Sharp-Ears* posits a comedy of manners within the orbit of fairytale opera, informed throughout by Janáček's most exuberantly folk-rhythmic score, with short dance strains played, repeated once, and retired only to turn up a few measures later in some modification. As with Wagner and Schoenberg (in *Gurrelieder*), the tide of fortune in nature's permanent regime is enhanced by a cyclic motival structure, not to mention animal buzzes and wails. But where Janáček's national predecessor Antonín Dvořák saw the forest as a jumping-off place for standard romantic sorcery and ill-starred love in *Ru-*

salka (1901), the more advanced composer seeks out its innocent cruelty, its inevitability. Compare, for example, *Rusalka*'s lovely but standard-make apostrophe to the moon, "Měsíčku na nebi hlubokém," to the monologue of Janáček's Forester in the final scene of *Sharp-Ears*, "Je to pohádka či pravda?," which passes from an idle reminiscence of the man's youth to a glowing comprehension of the inner passages of immortality. This same Forester falls asleep in the opera's first scene, and is awakened by a little frog who jumps onto his nose; in the finale, the Forester again drops off, so perhaps the entire opera is nothing but a dream—except this time, when the little frog jolts him awake again, it seems to be the first frog's grandson. "He t-t-told me about you," cries the wee mite, and the Forester, moved, stares about him in wonder.

After *Sharp-Ears,* Janáček soft-pedaled his instinct for lush string combinations, giving his last two operas a more angular and insistent quality. The bizarre waltz played over a bass in common time that closes Act I of *Věc Makropulos (The Makropulos Business,* 1926), the most brazen of Janáček's uncountable hemiolas, or the shrilling chains that clank in the prelude to *Z Mrtvého Domů (From the House of the Dead,* 1930), are among the few moments in this pair that address the average musical consumer who likes to buy his theatre prefitted, off the rack.

All these six Janáček operas are adaptations from fiction and drama, but the final two pitched on sources of unconventional promise for lyric drama. *The Makropulos Business* derives from— or, rather, is—Karel Čapek's play, which Janáček simply trimmed and scored, and a particularly conversational play it is, too. An incomprehensibly gnarled-up lawsuit detains the proceedings, but the central axis, revolving around the efforts of a 342-year-old yet eternally beautiful woman to recover the formula for the elixir that keeps her spry, inspired the composer to endow the musical stage with one of its most irresistible figures. The opera singer Emilia Marty, who has also lived as Eugenia Montez, Elsa Müller, Yekatyerina Myshkin, and Ellian MacGregor, is, at bottom, Elina Makropulos, the daughter of a seventeenth-century alchemist (who invented the elixir of life for old Rudolph II) and a woman of ruinous fascination—yet Janáček chose to limn her, in the imperiousness of decaying sensuality, only after he had cut himself off from the continuum of operatic vocalism. Not until the last moments of the piece, when Marty tosses the formula away and collapses, does Janáček give the voice its head and the audience its Big Tune. Fittingly, *From the House of the Dead* is even bleaker, for it borrows Dostoyefsky's autobiographical novel on life in a prison camp, and, also fittingly, the brutalized, subhuman existence of the inmates is filled out by scenes of oafish comedy. This is no satiric tragedy, however, no spectacle of puppets dangling helter-skelter on their rods, but it does mark the emancipation of de-genre-ized opera, gulling the pigeonholers with an individualistic tang.

Another emancipation is at work here as well, a quasiphilosophical one, for in leaving the "well-made" territory of bourgeois and peasant systems, Janáček gained the uneasy Graustark of the modern temper where the native pastimes run more to identity trauma than to love and death, and where the uttermost realism can only be expressed through fantasy. In *The Makropulos Business* and *From the House of the Dead*, Janáček addressed myths of modern art, those of the new quest into self, told in language less easily decoded than that of the old, for the modern worldview admits of no code at all, no system. Though located in modern-day Prague or a Russian prison camp, as here, operatic life is beginning to take on that hysterical edge of the satirist bedeviled, despite himself, with a romantic's idealism. This explains the clean cut, the transparency, of the music, but also its sometimes fulsome rhapsody. But how else can one assume two positions at once?

A very international figure, this Janáček, so far ahead of his countrymen that only recently has his output made its treaty with the non-Czech establishment. Admittedly, the Slavic branch of opera, by its linguistic distance, is doomed to remain largely a "national" preserve, whether it be such truly ethnic products as Jaromir Weinberger's *Švanda Dudak* (*Shvanda the Bagpiper*, 1927), a lively bit of old Bohemian legend, or this same Weinberger's *Lidé z Pokerflatu* (*The Outcasts of Poker Flat*, 1932), a bit of old Bret Harte. Alois Hába won some academic attention for his work with the quarter-tone; microtonality, however, makes his *Matka* (*The Mother*, 1931) a work of curiosity but of forbidding difficulty for singers, who are all too frequently off the mark by a quarter-tone as it is. Of Czechoslovakia's post-Janáček generation, the best known (though yet surreptitious where most houses are concerned) is Bohuslav Martinů, who occupies a point midway between Janáček's Slavic impressionism and the less restless figures of the French. In fact, more of Martinů's many operas were composed to French or English texts than in Czech, for he emigrated early on in his career, and in short order has swerved from the native flavor of *Voják a Tanečnice* (*The Soldier and the Dancer*, 1927) to the urbane *Les Vicissitudes de la Vie* (*The Vicissitudes of Life*, 1928), *Zeitoper* très continentale.

Martinů's most famous work is *Julietta* (1937), made from George Neveux's *Juliette ou la Clé des Songes*, a dream play about . . . well, this one rather defies synopsis. Surrealistic to the point of flying reality altogether, it takes a traveler into a paramnesiac seacoast town where blackout and déjà vu are what is being done, the vocal and orchestral components incorporating the metrical propulsions of Janáček under the influence of Debussy's classic amnesiac Mélisande.

Martinů dispensed eclecticism at will; he could borrow from any style and pay it off with wit, as in the dry buffa diversions in his *Comedy on the Bridge* (1937). Oddly enough, Martinů's most national opera would have to be *Ariane* (1961), again with Neveux; the nation, be it said, is France. *Ariane* is a *tragédie lyrique* with only Neveux's Freudian pen and Martinů's untrustworthy chordings, constantly slithering out of key, to distinguish it from the work of Lully and Quinault. If French opera has a strong folk examplar in its background, it has yet to be detected, but surely the literary-cum-historical precedents quoted whole in *Ariane*'s face and figure rate as France's own sort of national music drama—certainly enough national controversies were provoked whenever the form threatened to evolve under the influence of foreigners (foreigners, that is, other than Lully, who as the founder of French opera was granted exemption from having been born Italian). So, *Ariane* offers the sinfonia ritornellos, the da capo aria, the few words stretched out in repeats, the sturdy cut of jib about the tempos, and the full closes of a grander epoch than this, yet in the melodic and harmonic expression of the present and with the assistance of a minotaur who has obviously been catching up on his Jung. In this umpteenth re-creation of Ariadne and Theseus, set in Knossos, the beast of the maze cannily appears to Thesée as Thesée himself, thus to conquer—for what hero could bear to slay himself? Both Martinů and Neveux bring off their homage with grace, the most prime moment of all being the end of the first scene, when the marriage of "the princess and a stranger" is announced. "Who is the princess?" asks Thesée. The strain of the work's tiny overture breaks out briefly, and Ariane replies, in unaccompanied recitative, "I am. My name is Ariane. And you are the stranger. What is your name?" And the curtain immediately falls.

In nearby Poland, heading east now to titanic Mother Russia, nationalistic feeling about opera runs as high as in Czechoslovakia, but there is no Polish Janáček—not even a Polish Dvořák. There is a Polish Smetana, of sorts, Stanislav Moniuszko, whose *Halka* in 1847 holds the national regard accorded Smetana's *The Bartered Bride* in Bohemia, but one Polish composer, Karol Szymanowski, still partly unknown in the West, is to be compared to no one. More a modern eclectic than a folk-thrilled national, Szymanowski collaborated with the poet Jaroslaw Iwaskiewicz on the stupendous *Król Roger* (*King Roger*, 1926), one of the unsung masterworks of contemporary opera. For this look at classical order beset by pagan ecstasy in twelfth-century Palermo, the composer let monumental primitivism battle chromatic luxury just as King Roger battles Dionysus in Iwaskiewicz's libretto, adapted loosely from the tale of King Pentheus of Thebes and rather more loosely from history's King Roger. Not till the checkered stylistics of the fifties and sixties was such diversity of sound types to be heard again in opera, and yet no one has better mixed his materials than Szymanowski in this work, skirting the bitonal polyphony of Stravinsky but succumbing to the rhythmic concisions of Bartók and the piled fourths and orientalism of Skryabin. Later, Hans Werner Henze would borrow this Apollonian-Dionysian conflict for Euripidean psychognosis in *The Bassarids,* and Benjamin Britten utilized it for a contest of artistic policies in *Death in Venice;* in *King Roger,* however, the confrontation is one of humors, that of the West besieged by that of the East.

Not until the arrival of Kryzstof Penderecki in the 1960s did

Poland assert itself on international stages, and while Penderecki's *Die Teufel von Loudon* (*The Devils of Loudon,* 1969)
has made the rounds, as it were, its value as musical opera has
been contested. Based on John Whiting's play, Penderecki's own
libretto cuts too slim a figure in short scenes not much amplified
by Penderecki's score, yet one can hardly deny the growing
power of such few lyrical moments as the composer allows. This
austerity must surely be deliberate, for the man who composed
the stunningly expressive *Passion According to Saint Luke* is not
one to lack musico-dramatic resources. Thus the balloon-idyll of
hopaks and polkas as a fixture of Slavic opera is punctured by a
British play drawn from French history and turned into a Polish
opera in German, just as contemporary opera in general tends
now to charge across frontiers in an ecumenical rendering of tradition for polyglot utterance. The French episode in question
concerns the arrogant Urbain Grandier, a priest loved by a
hunchbacked prioress and then, when he rejects her, denounced
by her for witchcraft. Grandier's ecclesiastic enemies have been
waiting for a chance to destroy the prelate; this they do in repel-
lent physical detail. Most likely Penderecki chose his subject to
confront unflinchingly human bestiality within the framework
of operatic polemic, the von Einem-Henze school of composers
that attempts to politicize the spectator with antitotalitarian material. (Henze actually has shown himself eager to substitute a
totalitarianism of socialist youth for older so-called tyrannies;
we'll get to him in good time).

And so, speaking of politics, to Russia, Slavic and certainly
national but not so confined to quarters as the Czechs, Slovaks,
and Poles because of the reputations of Prokofyef and Shostakovitch and also, perhaps, because of the attention accorded to it as
a major world power. Russian opera stands at the core of the na-

tional school, for while it is at present less oriented to the literary nationalism of England and the folkloric musical syntheses of the U.S., it holds the patent on unadulterated folk pageantry, having established the folk-as-hero back in the late nineteenth century, achieving an apotheosis of sorts in Rimsky-Korsakof's mystical *Kityezh*.

Admittedly, nothing could be more adulterated—that is, conceptualized—than Rimsky's epic of patriotic trauma and pagan-Christian redemption in the trinity of Kityezhes, and his savagely satiric *The Golden Cock* both commends and yet improves the Russian zeal for getting Alyexandr Pushkin's work off the page and onto the boards, with a twist. But in such an entry as *Skazka o Tsarye Saltanye* (*The Tale of Tsar Saltan*, 1900), Rimsky set the bowstrings for the flight of the straight-arrow folkloric presentation, legend rather than myth, in a musical framework that is still seeing national service, though in far lesser hands.

For Rimsky-Korsakof's pupil Igor Stravinsky, however, only the new would serve. In his first opera, the one-hour *Solovyey* (*The Nightingale*, 1914), first produced in Paris as *Le Rossignol*, the young composer had already launched his own voyage. A time lag in composition between the first little act and the rest of the piece did allow for some Rimskyan imprint on the opening scenes, but Acts II and III leap the frontier, Russian still in voice yet cosmopolitan in technique. Styepan Mitusof's libretto on the miraculous nightingale and the Emperor of China was drawn not from Pushkin but from Hans Christian Andersen, and a healthy infusion of sardonic chinoiserie dispels any sense of folk location. The Nightingale is Indo-European in tone, not Russian (and certainly not Chinese), delicately evocative of the international storybook, apprehensive of international musical develop-

ment as well as tradition, and, at times—as when the ladies of the
Chinese court try to imitate the nightingale by gargling on a
fourth over a trilling orchestra—ineffably weird.

Like many another foreigner . . . like Lully, Pergolesi,
Gluck, Grétry, Rossini, Meyerbeer, Wagner, Martinů . . .
Stravinsky had come to Paris to make his name, and other Russians, after 1917, were to follow, although official censorship was
not to enslave the musical community until the party directives
of 1932, at which time so many individualistic composers were
either forced into mediocrity or outlawed that it becomes useless
to elaborate on "developments" in opera east of Smolyensk.
There was an understandable upsurge in revolutionary operas in
the 1920s in Russia—what the Nazis called "cultural Bolshevism"—and of course the dutiful setting of a libretto for a "people's opera" doesn't absolutely have to bring out the worst in a
musician, for Soviet life is filled with surface accommodations to
fiat. Gerald Abraham relates that *Tosca* was heard in Leningrad
at this time as *Borba za Kommunu* (*The Fight for the Com-
mune*) and *Les Huguenots* reset in St. Petersburg in 1825, as
Dyekabristi (*The Decembrists*).

But the leading rebellious artist of musical Russia had trans-
planted himself to Paris. Syergyey Prokofyef composed his first
opera at the age of nine, and his third, *Igrok* (*The Gambler,
1929*), was no less precocious, a minor masterpiece written when
he was twenty-five. Based, not too closely, on Dostoyefsky's short
novel detailing his own temporary obsession for the tables of
chance, *The Gambler* circles around and then into the cyclone
of compulsion with all the wheezes of the mature Prokofyef—the
abrupt modal key changes, the clockwork ostinatos, the headlong, downward-driving string runs, the perverse melodic intervals—but only with *Lyubof k Tryem Apyelsinam* (*The Love for*

Three Oranges, 1921), written after but produced before *The Gambler,* did Prokofyef get a foothold in the abiding repertory. *The Gambler* is a mordant comedy-drama, a grotesque, but *The Love for Three Oranges,* after Gozzi, is nothing but comedy . . . and also a grotesque.

In fact, the piece may be taken as a definitive attestation of neoclassical satire, closely akin to Busoni's *Turandot* in its mating of faerie and comic naturalism. As in Busoni, the commedia characters are plucked up from the past to cut the old didoes, and romance triumphs, though here there is none of the miniature psychology retailed by Busoni. This is pure comedy: it just happens, falling off a log only to remount it and prance, king of the log and the *logos,* king of logic and words. Music, at base, is its subject.

Prokofyef had left roiling Russia for the West in 1918, just when he began composition of his Gozzi opera, so the result was beholden to no particular social or historical directive, as Prokofyef's later works were. Whether it is because art doesn't take direction from outsiders or because grotesquerie suited the composer's autograph anyway, *The Love for Three Oranges* is his operatic trump card, daring less than the mammoth *War and Peace* but, in its own terms, fulfilling said terms with less qualification. The plot itself is framed by the antics of five groups of theatre buffs: the crackpots, the empty-heads, and the lovers of tragedy, comedy, and romance. The empty-heads call for "captivating nonsense," the last three for, respectively, "Grief! Decline! The suffering of fathers!," "Bracing laughter," and "Tender kisses." After a minor riot, the crackpots prevail, and it is they who present the entertainment proper, occasionally poking their heads out during the action to size up the situation or chase one of the rival groups back into the wings. As it happens, however, the show that the crackpots put on answers all the above demands. In it is a little something for everyone, for modern comedy is just that sort of hybrid, breeding laughter out of disaster as well as farce—if Weill and Brecht could write an opera for beggars, why not an aesthetic for crackpots?

So: crackpot opera for everybody. Lacking the moralistic commentary of *The Golden Cock,* it nonetheless offers its deputies of darkness and its magic devices. The Prince of Clubs suffers an apparently incurable depression, unexpectedly broken when the evil Fata Morgana takes a sudden tumble at a court pageant. The Prince laughs, in quite some detail, at the sorceress' indignity, and she in revenge enchants him with an obsessive passion for a certain three oranges, which he seeks and finds, but unfortunately in the middle of a desert. Out of each of the first and second oranges pops a young woman who promptly dies of thirst, but the third girl is saved by the crackpots, who provide a pail of water, and at length Morgana and her associates are vanquished and all ends joyously.

The composer wrote his own libretto—one of the most lithe and delightful of its kind—offering meanwhile some little treasure of invention on virtually every page of the score. For example, the transition from the prologue of the rabid dramaphiles to storybookland is a study in dramatic momentum: shushing the backers of other genres, the crackpots announce the title of their piece to a suitably churning accompaniment in the strings, which continues under a fanfare played onstage, all on one note, by a bass trombone. Suddenly, the music turns mysterious as a Herald takes the stage to proclaim the sad state of affairs at the court of the King of Clubs, whose son, he explains, ails from hypochondria (ah, yes: the suffering of fathers). Another salute from the trombone, this time with the assistance of a bittersweet motive on the violins, and both orchestra and crackpots prepare for the play with abandoned momentousness. "It begins!" cry the crackpots, followed by a martial tutti as the game of grief, bracing laughter, and tender kisses commences.

A national opera, of whatever sort, needs its national composers, and Prokofyef finally decided to face the *mouzhik* and return home permanently in 1932. In his Soviet entry, *Syemyon Kotko* (1940), set in 1918 in the Ukraine, where Communists battle loyalists and Germans, he aroused a critical tempest in the press, but glory as well as mixed reviews was his after the stage debut of *Voina i Mir* (*War and Peace*) in 1946. As in *Syemyon Kotko*, much of the mordant Prokofyef of his salad days was laid aside for patriotic salute, and the parallel between besieged Russia of the Napoleonic wars and that of World War II must have been overwhelming to *War and Peace*'s first spectators.

Tolstoy's novel is, it's true, too grand to settle comfortably between the proscenium arches in *any* form, and Prokofyef's opera exists in several editions, depending upon how much of the reduction and which revision one is willing to tackle in one night. Even to use every note would be to lose the book's transcendent development of time as a series of unperceived climacterics: such an effect, built up to in the graphic fullness of letters, cannot possibly be duplicated in opera, which by its every function throws these unperceived moments into the realm of Joycean epiphany. But, given that "flaw," *War and Peace* is a tremendous achievement, fast becoming a repertory item in the West for the dynamism of its self-convinced arioso and smashing spectacle to test the aptitude of singing-actors and technical crew. One of the few recent examples of the folk-as-hero pieces to be heard in the free world, *War and Peace* cannot fail to amaze with its collation of the personal and the epochal, even if the two admittedly were not serried all that skillfully. There is General Kutuzof, as big as war, chanting his—thus Russia's—inevitability through war to peace, or the touching death of Prince Andryey, or the massive final chorus, a frieze of the national will. (After checking over Prokofyef's piano score, Communist officials recommended that he emphasize the patriotic side of the story, which explains the sudden switch in focus from the personal characters, peace, to the epochal, war, in the eighth scene.)

Hemmed in as he was by the Ministry of Culture on one side and his responsibility to Tolstoy on the other, Prokofyef could not possibly have written his utter masterpiece this time out, but he came close. His own favorite of his operas—for a while at least—was *Ognyeniy Angyel* (*The Flaming Angel*), composed from 1919 to 1927 but not staged until 1955. A reasonably familiar item in German houses, this look into occult possession under the Inquisition includes appearances by Faust and the Devil, two standbys of the Teutonic imagination, and while Renata, the heroine who believes herself the lover of a flaming angel, carries the evening through the conviction of her music, Prokofyef eventually came around to believing most strongly in *War and Peace*, less tautly construed than *The Love for Three Oranges* or *The Flaming Angel*, but of a glorious, elemental breadth.

Taking a breather from national artwork, Prokofyef sneaked back into his sense of humor for *Obrucheniye v Monastirye* (*The Bethrothal in a Monastery*, 1946), a re-tailoring of Sheridan's *The Duenna* and a happy recovery of the fertile wit that had sparked so many sonata-allegro movements in symphonies, not to mention the Gozzi libretto and even the ballroom music in the second scene of *War and Peace*. This is sadly not the case in *Povyest o Nastoyashchem Chelovyekye* (*A Story About a Real*

Man, 1960), a very Soviet piece on the crippling, breakdown, and spiritual resurrection of a flyer who becomes a "real man"— a true Soviet citizen—by the application of a little strength through joy. The diatonic simplicity demanded by the Party clearly hampers even Prokofyef, the master of diatonicism—but then his C Major is not theirs. The work unfolds in thirty-nine scenelets, each a separate movement, as in *Wozzeck,* but the practice is evolved not from Berg but from Prokofyef's work with Eisenstein on the films *Alyeksandr Nyefsky* and *Ivan the Terrible;* it is the technology of the cinema, not of expressionism. Alas, too much of the piece is barely adequate and little of it penetrating—except for the valiant folk chorale that opens and closes the work—and no doubt one has the Central Committee to thank for an atrocious Slavic rhumba for a soldier's dance in the sanatorium episode.

The nullifying effect of censorship dogs the career of Dmitri Shostakovitch more than that of Prokofyef, for the "socialist realism" required of operatic composition by the Central Committee as of 1932 calls, among other things, for the simpleminded musical structure that makes opera operable for the masses, but which militates against the ambiguity, irony, and individual comity of art. This artistic determinism produces huzzahs for bureaucrat-composers of no talent but so complicates the creative procedure of a Shostakovitch that it is difficult to discuss Soviet opera except to state that it exists and to pass on to real worlds. Forbidding the practice of "formalism" (here meaning not the use of classical structures and developments, but rather a "decadent" contemporaneity of no utilitarian function—i.e., of no propaganda value), the Party has made its national music equivalent to the mass media of the free societies, organs of debased, toothless attitudinizing for which the hack creator is exclusively

suited. Still, Communism in general aims at just this sort of tax-
ing the fleet to soothe the slow, and given that premise, the present national opera of Russia has succeeded admirably in ac-
complishing its goal. It is the most sorry collection of lumpen,
dispirited carcasses of operatic form in the Indo-European world.

For that reason, Shostakovitch must be forgiven his banal musical comedy about the housing shortage in Moscow, *Moskva,
Cheryomushki* (*Moscow, the Cherrytrees,* 1959), as something
he most likely would not have had to bother with in a free country, but what is one to make of *Katyerina Izmailovna,* the 1963
revision of his *Lyedy Macbyet Mtsenskowo Uyezda* (*Lady Macbeth of the Mtzensk District*)? Damned in the pages of *Pravda*
two years after its premiere in 1934, *Lady Macbeth* retold Nikolai
Lyeskof's story of a discontented wife undone by love for one of
her husband's workers. It may sound like socialist realism, but
apparently when Shostakovitch finished setting A. G. Pryeys' libretto, it wasn't. For socialist realism, Stalin and his protégé in
the Politburo, the gigantically ignorant Andryey Zhdanof, preferred Ivan Dzerzhinsky's *Tikhy Don* (*The Quiet Don,* 1935),
an adaptation of Sholokhof's famous novel, and though work-
manlike, without the violent intensity with which Sholokhof
defined his Cossacks.

According to the official essay in the state-published piano-
vocal score of *Katyerina Ismailovna,* Shostakovitch's second version, among other things, delivered the work from certain "eccentricities"; given Shostakovitch's style, that must have been
quite a job. But how are we to know, finally, how much of this
Katyerina Ismailovna is an author's honest revision and how
much a prisoner's remission? Poor judgments abound in the
piece as today performed, as in the obtuse ballroom waltz that
intrudes on the father-in-law's monologue in Act II, and much

of the score is too glib, too restively "operatic," with dutiful gestures of tone- and metre-change at the entry of each new speaker. On the other hand, Shostakovitch was one of the century's outstanding composers, rather underrated at present, and even a failure by a man such as this must bequeath its treasures, among which would be the drastic passacaglia interlude between the two scenes of Act II, a touch of "formalism" too expedient to the idea of music drama to be decried even by *Pravda*.

That Shostakovitch should have done less than his best in naturalistic melodrama is all the sorrier when viewed in the light of his so very Russian relish for fantastic satire; his wonderful requisition of same for *Nos* (*The Nose*, 1930), from Gogol's short story, proves what potential he had to grow old with in opera. Indeed, Shostakovitch, as much as anyone—and more than most—displays the change of tone in the serious music of the postclassical age, when hollow grotesquerie rises to the pinnacle and the age of Godot commences—the looking around, the waiting, the anarchic pranks and pratfalls, the religious nihilism. Like a death rattle, the woodblock and strings *col legno* of the prison sequence, "F Tyurmye," in Shostakovitch's Fourteenth Symphony, sound the empty new code for handling what Beethoven in the opening of *Fidelio's* second act could not bring himself to score without some arc of heroistic hopefulness. Romanticism, however, is officially over. Isn't it?

The Nose, prime early-modern farce (Arthur Jacobs calls it "the comic *Wozzeck*"), clearly points the way. Passages of it reek of the contrapuntal faux pas and in-jokes of *Zeitoper* and *Jazzspiel*, but the whole adds up to a vital spree of angry, not lighthearted, comedy. A. G. Pryeys penned this libretto, too, for the composer, making it something of a literary jaunt by gleaning scads of quotations from not only *The Nose* but other of Gogol's works (plus a bit of *The Brothers Karamazof*), and while conservatives fumed at this disrespectful chowder of classics, the result blends into a transformation of material, faithful to its source in spirit but its own item in statement. The nose of the title has vacated the face of one Major Kovalyof and goes on a rampage of antisocial behavior, even unto impersonating a state official, until it is arrested and remanded in poor Kovalyof's custody. Throughout, the unscrupulous non serviam of fantasy is mirrored in the music—in the use of that insufferable musical saw, the flexatone (fresh from its stint in Schoenberg's Variations for Orchestra); in an entr'acte for the percussion department; in the Police Inspector's impossibly high tenor part (higher even than that of the Astrologer in *The Golden Cock*); in an eight-part fugue for the simultaneous dictation of eight advertisements in a newspaper office scene; in the offstage chorus that ushers in the disguised nose, who/which sings at its former owner in a nasal "French" tenor. With such wonders to work in the age that called, more and more, for the mongrel *entrechat* of the bizarre, the loss of Shostakovitch to Party hoodoo is one of opera's tragedies.

One work that triumphed over official interference, perhaps because its author's views coincide with the Party line, was Yuri Shaporin's *Dyekabristi* (*The Decembrists*, 1953), on the unfortunate revolt of 1825, and a somewhat Meyerbeerian spectacle that builds to a dramatic reproduction of a historical disaster; the opera itself built up to a triumphant production after several decades of composition. On the far end of the pole from historical re-creation is Dmitri Kabalyefsky's *Mastyer iz Klamsi* (*The Artisan of Clamecy*, 1938), known generally as *Colas Breugnon*, and based on Romain Rolland's novel rather than the more usual Russian source. It is, certainly, a curious choice for Russian opera. Colas, a sculptor who brightens up sixteenth-century Burgundy with wit and genius, is clearly not going to demand a *Nose*-like black romp or a romantic-historical *Decembrists*-epic, but Kabalefsky's music nonetheless would do for any Russian narrative, showing its roots almost in direct line of growth from the Glinka of *Ruslan and Lyudmila* and waxing especially Slavic when the action is supposedly at its most French. A revision of the work in 1968 was followed by the indispensable Soviet piece, *Syostri* (*The Sisters,* 1969), a dull evening devoted to a pair of sisters who dream of becoming navigators. No conflicts are placed in their path, and they become navigators. Enough said.

It is in its synoptic collectivism that modern Russian opera shows itself most backward—and yet most traditional, for the scope of the folk-as-hero was originated back in the days of Mussorksky and Borodin. Though Russian audiences can enjoy the life and times of the individualistic Colas Breugnon, as they did those of Chaikofsky's Yevgyeni Onyegin, for example, their view of opera as a vehicle for national myth—for the alleluia of folk—is older than socialist realism. The insistence on simplistic musicianship, however, is nothing less than murder, and the belief in this protagonism of the people locks the librettos in a phoney Oz just when Western opera is most concerned with that realer Oz of personal identity.

Having dealt with Slavic opera at some selective length, and with other ethnic groups yet to be considered, each in its custom, we ought to confront this question of artistic nationalism, for now, circa 1935, is the ground-zero of patriotic exhibit in the West, when a world war and a world depression threw off the protective coalition of this region and that for the determinism of isolation, and when central Europe's totalitarian governments promised further trouble. This would have been a likely heyday for national art, but the jarring rise of neoclassicism, with its off-putting structuralism, its twelve sullen tones, and heedless parody, proposed new tenets for musicians to debate in crossfire, and the emphasis now will be more on form than on tradition—on *how* it is to be worked out rather than how it had been worked before. Theatre and fiction were, of course, ahead of the game; they always are. Georg Büchner, Anton Chekhof, and Alfred Jarry all arrived before Debussy's *Pelléas et Mélisande* and Bartók's *Duke Bluebeard's Castle,* and James Joyce nearly erases them all topping the profane with the sacred, realism with myth, in *Ulysses,* a book. But opera comes along into the future in its own good time, and much of its questing flux in the twentieth century was bound, finally, to break with the subject matter that had been too useful too long. Just as Janáček was graduated from the household, love-and-death realism of *Jenůfa* to less "operatic" material in *The Makropulos Business* and *From the House of the Dead,* and just as Stravinsky abandoned the *Märchenoper* of *The Nightingale* for commedia and tragedy in *Renard* and *Oedipus Rex,* so did their colleagues seek out new topics, and so did the national fortes have to assert themselves in style and manner up-to-date. Here is where the literary opera, radiant with the prescience of poets, and the historical study, come into their own. Out with the dirndl and the strophic chorus; they've said as much as they're ever going to.

We've already noted, in passing, that Carl Nielsen adapted a comedy by the classic Norwegian-Danish dramatist Holberg for *his* splurge in national opera, *Maskarade,* and Nielsen's only other opera, *Saul og David* (*Saul and David,* 1902), devolved on religious, not folk, legend. Similarly, the Swedish repertory is best known not for folk works but for the avant-garde individualism of Karl-Birger Blomdahl and Lars Johann Werle. Werle's opera-in-the-round, *Drömmen om Thérèse* (*Dreaming About Therese,* 1964), uses prerecorded electronic sound along with speech and highly advanced song in an adaptation of Emile Zola's story, "Pour une Nuit d'Amour." There is some parallel between Zola's plot and Strindberg's *Miss Julie,* but the point of the work is its aural-spatial experiment, and even amidst the nineteenth-century costuming *Drömmen om Thérèse* is "something completely different." Blomdahl's famous *Aniara* (1959), the science-fiction opera about an international spaceship running helplessly off its course, is in its dodecaphonic collage about as unnatural as opera can get; and Blomdahl tacked to the literary in *Herr von Hancken* (1965), based on Hjalmar Bergman's novel, and while definitely Swedish in language, definitely not for the folk. Erik Lindegren's libretto is lavish in symbolic wordplay, picturing the adventures of a Swedish captain and his family at a health resort, where Death and the Devil shrive the ridiculous captain of his vanity to sounds of post-Schoenbergian nullity.

Less the literary than the historical rules the Argentinian Alberto Ginastera's operas, *Don Rodrigo* (1964), *Bomarzo* (1967), and *Beatrix Cenci* (1971). All three are serial works—could there ever be a serial folk opera? No, never—but the first is less so than the others, employing the baroque palette favored by Penderecki to keep the spectator at bay. Hisses, shrieks, moans, bells, and cannon, entreating offstage choruses give such operas pace if not true melody, and a good production of *Don Rodrigo* can always skate by as a "theatre piece," even if the omnifarious sound effects give the impression that the work was composed by a kaleidoscope. *Don Rodrigo,* the exploits of an eighth-century Spanish Visigoth king, is bourgeois atonalism, acceptable to conservatives, but *Bomarzo* and *Beatrix Cenci* have proved harder to enjoy, despite stormily sensual librettos, for here the tone clusters and aleatory happenstance of the true modernist take the reins; on the word side, *Bomarzo* is the more interesting, made up of the last hours of a hunchbacked Italian duke, reviewing his ugly life in flashbacks as he slowly dies of poisoning.

One purveyor of literary opera in Spanish, Manuel de Falla, turned almost immediately from the exactness of his page of Cervantes in *El Retablo de Maese Pedro* to the span of myth in *L'Atlántida* (1962), a "scenic cantata" on the epic poem of Mossén Jacinto Verdaguer. The composer left the work in some disarray at his death to be finished by his pupil Ernesto Halffter, and, term for term, this is no opera, but if the text is deferred to chorus and choral soloists rather than characters, the action is staged (for dancers and actors) and the action is, ultimately, musical. Starting more or less at the creation of the world and carrying on through the glory of the Spanish empire, *L'Atlántida* is one of the few epics of undiluted neoclassicism, a bizarre reconciliation of pagan rumor and Christian chronicle; within its cubits, de Falla made a strong case for the quasianonymous popular entity, one of the last before the Herr von Hanckens and Don Rodrigos usurped the stage.

One racial group has yet to be heard from, and they have been saved for last because they are the most remote of all the so-called nationals, the Finno-Ugric peoples. For if Czech and Russian are out of the ken of the Germano-Romantic opera world, surely Finnish and Hungarian are hopelessly beyond reach. Virtually no Finnish music drama has been granted a hearing on any stage outside Scandinavia, and of the Hungarian college little is taught beyond *Duke Bluebeard's Castle*.

As it happens, neither the Finns nor the Hungarians are still trapped in that cottage industry, folk opera—remember what a grandmaster's gambit of contemporaneity Bartók's opera had been earlier on, even with its folkloric, "ancient eight" scansion. True, there has been the considerably folkish Zoltan Kodály and his musical comedies *Háry János* (1926) and *Székelyfonó* (*The Szekely Family's Spinning Room*, 1932)—*very* national, complete with cimbalom—but of late Sándor Szokolay has disarmed the ready vernacular of the merry villager for modern individualism in adaptations of Lorca (*Vérnász*, 1964, from *Blood Wedding*) and Shakespeare (*Hamlet*, 1968), each faithful in spirit to the source and each to be compared favorably to competitive versions by Wolfgang Fortner of the Lorca and Humphrey Searle of the Shakespeare.

As for Finland, its opera production is the most secluded of the major European countries; unlike the odd mounting of *Háry János, Maskarade*, or *Król Roger* in central Europe—Germany, usually—Finnish operas have not, so far, traveled. All the same, there is no shortage of opera composition at home, particularly of the emancipated modern brand that prefers personal drama to national parade, though a work such as Aulis Sallinen's *Ratsumies* (*The Horseman*, 1975) does not fail to apostrophize the ongoing imagism of Finnish feeling in much the same way that Rimsky-Korsakof's *Kityezh* does for Russia, saving Rimsky's green world symbolism. Intelligent of both contemporary musical philosophies and contemporary textual commentary, Finnish opera has developed in the last few decades from an essentially backwater diversion into an impressively genuine modern theatre, looking in on early Freudian analysis in Tauno Marttinen's *Poltettu Oranssi* (*Burnt Orange*, 1971), or on the question of art versus pop in Einojuhani Rautavaara's *Apollo and Marsyas* (1973), which updates that famous musical tourney with quotations from the old masters of the symphonic era and jibes at the new masters of the age of rock.

Modern opera synthesizes its stories for soloists out of the accrued achievements of the global community, making ultraindividualism out of the information of the races—Busenello, Lully, Pergolesi, Gluck, Favart, Marschner, Scribe, Wagner, and Verdi had all put a particular stamp on the dies even before *Pelléas et Mélisande* overhauled method, and now opera carries no one passport. Still, there is nothing like seeing a native company honor its native art with that facility and apprehension that foreigners will always lose in translation. See, for example, an English troupe bite into *Albert Herring*, or real-life Czechs send up *Mr. Broucek* on its two excursions, or the Bolshoi Opera's astonishing production of *The Gambler*, played on a unit set that revolves, at the finale, like a giant roulette wheel—this is the study of style.

Obversely, style may adapt to style, as when *Blood Wedding* or *Hamlet* turn into quite acceptable Hungarian operas, or when the Italian composer Roberto Hazon fashions *Una Donna Uccisa con Dolcezza* (1967) from Thomas Haywood's Elizabethan play, *A Woman Killed With Kindness,* via the lithe duos and tempo di blues of the Menotti era, or even when the English film comedy *The Ladykillers* turns up on the stages of Czechoslovakia as Ilja Hurník's *Dama a Lupici* (*The Woman and the Thieves,* 1974), lovable old eccentric, avuncular bobbies, and overgrown T-boys all compact.

VI

The Heroes
of Good Adventure;
or,
The Masks Are Down

I. FRANCE

One knows, positively, that the bimodal medium of satiric tragedy is in the ascendant when a Frenchman presents the story of St. Joan as a musical *mystère-cauchemar*, with a Joan who only speaks, an assault of medieval pastiche and modern arioso, a heresy trial presided over by a pig and attended by baaing sheep, and a distinctly understated heavenly choir. The words by Paul Claudel, Arthur Honegger's *Jeanne D'Arc au Bûcher* (*Joan of Arc at the Stake*, 1938) revamped the sacred instincts of Debussy's *Le Martyre de Saint-Sébastien* for the indecorum of new times, enlisting its sacrosanct moral tone while emphasizing the intrusion of the speaking voice. Nothing of its day is more vulgar and barbaric and funny than Joan's trial; nothing more inspiring than Joan's reminiscence of her childhood and the sweetness of spring in the forest. Claudel, of course, was no on-and-off hackeur of instant librettos, but a playwright and poet who would infuse opera, the least innovative of the arts, with the vivacity of the literary stage, and for this entry Honegger was willing to elaborate on the austerity of his *Antigone* style, learning parody.

The French had been there early on with the simple parody of spoof in the old ballad comedies, but modern parody calls for severed slapsticks, the two-pronged weapon that both borrows and cultivates. It is not enough to drop peasant rondelays into a maytime sequence and say *Voilà*: one must connect the associative derivation of the borrowing with the irony of dramatic con-

text. What, for example, does the simplicity of childhood folklore mean to a woman about to be burned at the stake after a
short but reckless career as a soldier? What effect does this folkish diatonicism have on the audience after the animal grunts and
mewing of the trial scene? And, particularly, what reference does
it make to its time, western Europe in 1938?

As parody is a parcel of the neoclassical delivery so welcome to
French music in the twenties and thirties, it was let off the leash
to do what it does in both serious and comic works, even as the
formal distinctions of the one and the other were starting to fade.
Parody's great benefactor in France was Darius Milhaud. Extremely catholic in his styles and stances, Milhaud has been more
the technical bravo than the passionate lyricist of instinct; he has
composed over fifteen operas of all sizes and intentions, but
neither in the bouffe nor in the elegy does his music quite take
on that land's end of rightness and marvel that even Massenet's
did. Milhaud is witty and sure, and never dull, but his superb is
fragile and his gravity often lightheaded. No one musician was
able to turn so many influences to so much use so efficiently, and
in his own neoclassicism he is never overtaken by influence, always the master. But he may have ridden too far on parody, for
audiences have found him as dry as tinder, and outside of France
he is seldom performed.

Still, Milhaud dominates the era, from *La Brebis Egarée* (*The
Strayed Sheep*, 1923) to *La Mère Coupable* (*The Guilty Mother*,
1966). Not unequal to monster opera, he is cited for the mammoth *Christophe Colomb*, but his heart must beat alla breve, for
it is in the shorter works that he most excelled. Just after finishing the three parts of the previously discussed *Orestie*—all angles
and planes, like an Ayn Rand hero—Milhaud grew tender in
Les Malheurs d'Orphée (*The Sorrows of Orpheus*, 1926), a

lightly scored and unpretentious pastoral in which Euridice does not return from hell: Orpheus doesn't attempt to retrieve her. Instead, Armand Lunel's libretto yields three village workers, an animal quartet, and Euridice's three vindictive sisters for dramatic initiative, with Orpheus a rustic pharmacist and Euridice a gypsy belle. Not a comedy, really, *Les Malheurs d'Orphée* is best described as a cute tragedy: the opening scene, when the three villagers come looking for Orpheus, is composed in the animated 3-3-2 rhythm over running eighth notes familiar from Latin American rhumba bands (Milhaud had spent World War I in Rio de Janeiro as secretary to Paul Claudel, France's minister to Brazil), and the final cataclysm, Orpheus' death, is precipitated by the arrival of Euridice's sisters, who serve as Bacchantes by tearing Orpheus apart with scissors, whip, and garrote, and who hum quietly while he dies.

Milhaud had visited Harlem, too, and jazz was another of the many wellsprings of his parodistic current, as were the singular cadences of his native region, Provence, and the dauntless step of the infamous Java. Unlike the masters of the romantic era, who tended to specialize in this or that format, Milhaud tried everything, and when *Zeitoper* flowered, his watercan was ready to tip. Pouring, it presented the "trois opéras minutes," three ten-minute pieces scaled down for satire and purity, *L'Enlèvement d'Europe* (*The Rape of Europe*, 1927), *L'Abandon d'Ariane* (*The Desertion of Ariadne*, 1928), and *La Délivrance de Thesée* (*The Salvation of Theseus*, 1928)—notice the classical subjects. The best of Milhaud's one-acts was an early work, *Le Pauvre Matelot* (*The Poor Sailor*, 1927), to Jean Cocteau's text about a sailor's wife, waiting faithfully for her husband for fifteen years, who is visited by a man who tells her that her husband is on his way and whom she murders for a pearl necklace;

the man is, of course, her husband. Subtitled a "complainte" (lamentation), *Le Pauvre Matelot* sounds somber throughout, yet in that off-putting music-versus-words manner to be expected of the interwar years, with the grip of bitonality and parallel ninth chords hectoring the senses.

If the Swiss Honegger could give the French a national sort of endeavor by hymning the passion of Saint Joan, the birthright Frenchman Milhaud proved ecumenical and sought out the New World or the Bible when flirting with the grand ceremonial. *Christophe Colomb* (1930), *Maximilien* (1932), *Bolivar* (1950), and *David* (1955) are his larger works, *Colomb* being the centrus of Milhaud's reputation as a serious composer of opera and an inadvertent summation of operatic neoclassical technique as it had been developed in the twenties. In this life story of Christopher Columbus, a narrator lends the status of heroic chronicle (accompanied by percussion out of *L'Orestie*), symbolism and expressionism regulate some of the scenes, a prominent chorus makes its commentary, and, o tempera!, newsreel footage adds to the debate by duplicating the staging in slyly alternative versions, repairing life's truths for glamor's sake, as any respectable newsreel does.

Paul Claudel's libretto for *Christophe Colomb* tended to religious allegory, but Milhaud never quite bought the total projection of the sacred—or maybe he simply could not achieve it in his scientific instinct for method. At the end of Act I, Columbus' men are threatening a mutiny, and if the onstage tension is mirrored in the music—which assumes seven different tonalities at once—this also provides its own separate technological tension, forcing even the layman to consider how the musician-scientist is going to solve the tonal puzzle. Thus does Claudel fall somewhat by the wayside; the musical resolution is a B Flat Major triad, o mores!, but meanwhile the mutiny, also resolved, has been submerged in neoclassical know-how.

Later in his career, Milhaud pulled back from neoclassicism somewhat. The *David* he wrote for an anniversary festival in Jerusalem, though as static as *Christophe Colomb,* showed less of the zeal for parody and more . . . well, heart. But for all that it remains far less interesting than the multiscoped *Colomb,* which shares with *Jonny Spielt Auf* and the Weill-Brecht pieces a sense of era in social as well as artistic terms. When last heard from, in 1966, Milhaud joined Mozart and Rossini in setting Beaumarchais' *Figaro* trilogy, in *La Mère Coupable (The Guilty Mother),* in which the Countess is the mother in question and naughty Cherubino the father. Arid! Superfluous! were the cries of the folk on hearing the work at its premiere in Geneva.

Parody suited all Frenchmen of the thirties, it seemed, and not surprisingly, since it was the French who for years ran the most thriving business in musical-comic wit in all of Europe, and who made the lasting contributions to musical satire in the nineteenth century. Admittedly, the fame of those who headed the Gallic musico-comic concern has tarnished with the years, and national comedy, like poetry, doesn't translate well enough for outlanders to know for sure. Henri Sauguet, Charles Levadé, Jean Rivier, Marcel Samuel-Rousseau, Marcel Delannoy, and Jacques Ibert are the names associated in the concern circa 1935; the titles themselves are all but evaporated. Delannoy's works are occasionally heard in Europe, especially his gleaning from the *Decameron, Ginevra* (1939), a haunting score, and Ibert's *Angelique* (1927) turns up in one of the lesser halls in Paris every so often, as does Albert Roussel's *Le Testament de la Tante Caroline,* noted here earlier. But none of these minor masters, amidst the double entendres and saucy couplets that the libret-

tist Nino could devise for them, could equal the wit of Ravel's
L'Heure Espagnole, and soon enough it was clear that French
comic opera had hit a snag after Ravel's *L'Enfant et les Sortilèges* in 1925, as airy as *L'Heure Espagnole* had been earthy, but
no less keen.

Ibert, at least, recanted in *Persée et Andromède* (1929), in
collaboration with the unstoppable Nino, this time in an *opéra-
ballet* in the fashion of Lully. Ibert teased the seriousness of the
occasion by letting Perseus woo his princess with a Spanish serenade, but otherwise the piece is polite, if not exactly reverent
("Et maintenant, la belle," says Perseus to Andromeda after he
slays the monster, "Hop! A Cythère!"), closing with a beauty-
and-the-beast twist: the monster turns out to be a prince in dis-
guise, and since Andromeda will have nothing to do with the
arrogant Perseus, she and the monster-turned-into-a-beau take
the happy ending. It's hard to figure exactly what Nino was getting at, as his libretto is neither spoof nor idyll, but Ibert did
score the verses with the most adult intentions, much of the
opera harkening back to the declamation-cum-impressionist-im-
provisation of the Debussy and Dukas Maeterlinck operas.

Not till 1947 did the energy of the *gai primitif* make a comeback, the circumstance being *Les Mamelles de Tirésias* (*The
Breasts of Tiresias,* 1947), Francis Poulenc's setting of Guillaume Apollinaire's absurd little play, a relic of early-middle
Dada. Poulenc had made his first stab at comic opera as early as
1920, but *Les Mamelles de Tirésias* signals his emergence, after
decades of suites and song cycles, as a stage dramatist of prowess.
Apollinaire's preposterous farce helped sound the revolution in
pre–World War I days, but as an opera it was à la mode in 1947,
sharp, quixotic, and very vocal. The heroine of the event is
Therese, a discontented wife who grows a beard, sends her
breasts into the air in the form of balloons and explodes them,

rebaptizes herself as Tiresias, and disappears while marplots of farce wrangle and cavort in puns and non sequiturs. The abandoned husband, knowing his patriotic duty, decides to go ahead and have children without his wife's help; a few hours later, he has birthed 40,049 babes. But now Tiresias returns as a fortune-teller, all in favor of family life and procreation. She and her husband lead the cast in assuring the spectators that they must have children, inspiring A Fat Lady and A Bearded Man in the audience to exhort their neighbors to do their part: message, cheers, finale.

Musically, Poulenc's individual sound gets a lot of speed out of altered chords and the 9-8 suspension, but his freshness of approach in comedy derives not only from imbecile jingles but from judicious application of a genuine lyrical gift, so that the serenity of Poulenc the songster is never too divorced from the savvy of Poulenc the joker. Thus, a calm interlude in the fracas is provided by Therese, her husband, and the chorus indulging in wordplay, this nonsense set forth in a supple hymn that moves from simple C Major into the lush wonderland of $\frac{7}{6}$ chords:

Comme il perdait au zanzibar,
Monsieur Presto a perdu son pari
Puisque nous sommes à Paris;
Monsieur Lacouf n'a rien gagné,
Puisque la scène se passe à Zanzibar
Autant que la Seine passe à Paris.

As he lost while playing zanzibar,
Monsieur Presto forfeits his bet
Since we're in Paris;
Monsieur Lacouf has won nothing
Since the scene takes place in Zanzibar,
Just as the Seine passes through Paris.

Lacouf and Presto, in fact, have just quarreled for no particular reason and have shot each other in a duel; their bodies litter the stage, but farce-life goes on in music of pastoral beauty. This was Poulenc's gift to French comedy, a far greater one than those lavished on Nino's librettos by Roussel and Ibert.

But meanwhile, the question was: who would take up where Debussy and Dukas had left off, and, for that matter, where was the next Massenet? French opera, not just French comedy, suffered badly after the twenties, and only one work since then has achieved international status, *Dialogues des Carmelites* (*Dialogues of the Carmelites*, 1957), also by Poulenc, this one his sole full-length serious opera. Perhaps it is some measure of the collapse of the national haunt, l'Opéra, that *Dialogues des Carmelites* had its world premiere at La Scala in Italian translation, though when the Palais Garnier mounted its production five months later, it did fill the stage with an imposing collection of grandes dames—Denise Duval, Régine Crespin, and Rita Gorr shared the foreground—and the piece's quick traversal of the Western circuit, not excluding an afternoon on American television, took up some of the slack left by vanishing French musical leadership.

In this setting of George Bernanos' play about a noblewoman who becomes a nun and faces martyrdom with her sisters during the Reign of Terror of the French Revolution, Poulenc omitted an opening crowd scene showing the rabble rehearsing for mob justice, and many an internal exchange to boot, but what remains is a striking example of the nonoperatic, nonlibrettistic libretto that works effectively as opera. These are, absolutely, dialogues of the Carmelites: the Prioress instructing the heroine, Blanche, in the rigors of vocation; Blanche probing the innocent, disquieting chatter of Sister Constance; the sisters coping with the Prioress' ungainly death; the new Prioress addressing her charges; Blanche receiving her brother and refusing to leave the convent; revolutionary officers expelling the sisters from their home; the sisters voting on whether or not to take an oath of martyrdom; the new Prioress again addressing her flock—this time in prison . . . and a last dialogue, this one with the Virgin, sung on the scaffold as the soprano choir, punctuated by the fall of the blade, is subtracted by one at a time.

A wordy but dramatic work, with perhaps the most affecting final scene in all French opera, *Dialogues des Carmelites* bears the marks of national tradition in ways that *Pelléas et Mélisande*, an opera much like it and yet not like it at all, does not. *Pelléas*, all light and shadow, all nature play subtly drawn, cut the Gordian knot of word and music; *Dialogues des Carmelites* gathers up the loose ends in the tensile, symmetrical phrases of French theatre music going back to Lully and Gluck, veneered in the unmistakable strain of baroque *opéra-ballet*, as in the striking first measures of Act III. The ample chording of Poulenc's harmonies gives the Carmelites' arioso a beauty to mitigate the austerity, and altogether the work has a shape and presence one would not have expected from Poulenc, the miniaturist voluptuary.

Unfortunately, arioso gave way to lackluster recitative in Poulenc's last opera, *La Voix Humaine* (*The Human Voice*, 1959), composed so Denise Duval could hyperventilate in Jean Cocteau's abysmal playlet about a woman's final telephone conversation with her departing lover. An unsettling delusion of adequacy permeates the work's endless forty minutes; scarcely a moment of real musicality invades the moaning and mooing until at last the unseen lover rings off, and Poulenc, inexplicably, quotes the fate motive from Wagner's *Ring* and draws the curtain.

But, still, who was to take up Massenet's mantle, who provide the zesty musical romances that had kept *Le Jongleur de Notre-Dame* and *Don Quichotte* floating in their day? Like Massenet, the young hopefuls were devout adapters. André Gide sang in Guillaume Landré's *La Symphonie Pastorale* (*The Pastoral Symphony*, 1968), Alfred de Musset in Daniel-Lesur's *Andrea del Sarto* (1969), García Lorca in Louis Sauguer's *Marina Pineda* (1970), Jean Anouilh in Jean-Michel Damase's *Eurydice* (1972). With the exception of the scathingly evocative *Andrea del Sarto*, the idiom was conservative, both in textual format and tonal stance, as if, unhappy with atonality and bored with the wacky isms of the twenties, French composers were in alignment to the near side of Poulenc, deliberately trying to keep the conventions going and the people appeased.

Easily the most successful such was *Opéra d'Aran* (1962), composed by Gilbert Bécaud, the pop tunesmith and cabaret diseur of "Laissez Faire, Laissez Dire" and "Je Veux te Dire Adieu"—and lo, the people were appeased. Bécaud's piece is old-fashioned but undeniably theatrical, recounting the visit by an Italian dreamer to a barren Irish island where he deposits equal parts of poetry, gaiety, and, opera being what opera is, tragedy. Actually, *Opéra d'Aran* is almost what a musical comedy would be if musical comedy were through-composed, with an atmospheric chorus of the islanders before the curtain rises ("Aran . . . proffered as an offering, rejected by the universe . . ."), a bloodcurdling death during a storm, the startling, wordless return of a long-vanished suitor to claim the soprano from the Italian, the accidental blinding of the soprano during a fight between the two men, and, the only way out, the soprano and the Italian pushing off in a boat to drown together. A touch of *Jenůfa* here, a hint of *Peter Grimes* there, and a bit of *Brigadoon* all around—but even so, melodramatic as it is, it works.

These are the line-holders, resisting the encroachments of formalism and unfriendly twelve-tone with melody and stories; each nation, as we shall see, has its resistance overground, though France, Italy, and the U.S. emphasize it. Bécaud in his one opera and his colleagues mentioned above in their several outings have each been confronting the moderns since the late forties, but recently another kind of avant garde has arisen to challenge them, the other moderns of antiopera. This is the sect that rejects both the tune and the tone (row), and both the well-made and the absurd libretto. No, for these revolutionaries nothing coherent, however difficult or intellectual, fulfills as art; they demand the new upheaval. Gluck, Wagner, and Schoenberg preceded them on the barricades, but if music drama has more rebellion coming, these sansculottes are not the men to lead it. Idealistic they may be, but they still get boos, thoroughly merited, for everything they do, for what they do is pointless, self-indulgent, and dull.

Just as we shall be meeting more of the conservatives anon, we shall be meeting more of the revolutionaries, so let one French example suffice, Henri Pousseur's *Votre Faust* (*Your Faust*, 1969), a "fantaisie variable genre opéra" built on a phoney premise of audience participation. The "variable" obtains in that the spectators are free to shout no at any moment, forcing the action to tack, or heel, or sit up and beg—but as improvised opera is hard to guarantee, Pousseur and his librettist Michel Butor have to prearrange the no's via plants in the house. Parody there is in plenty, of course, in quotation from potboilers of tradition, but unlike the synthesists of neoclassicism, Pousseur quotes to no purpose, trying to create a context out of the simple feat of quotation. With the small orchestra seated on stage, and the projections and modern dress, *Votre Faust* is a paragon of the "opera" medium that will be newborn or die in the attempt. "I will show you something different," said T. S. Eliot once. "I will show you fear in a handful of dust."

2. ITALY

While most of the Western opera-makers bowed to renewal after 1930, Italy concentrated on protracting the projects chosen for it in the early 1900s. Except in the careers of Malipiero and Pizzetti, the revelations of neoclassicism did not interfere with the consecution of late romantic stories and sounds, with the star-crossed lovers, the somewhat miscellaneous Big Tune reprised at moments of bravado, and the merry-villager choral back-up oiled and ready. As for the symphonic-opera composers who were fired by D'Annunzio's rash *eroi* and *innamorate,* their era had largely ended: Zandonai ceased operations in the thirties and Montemezzi in the early forties, only Alfano surviving World War II, with *Il Dottor Antonio* (1949).

Alfano composed *Cyrano de Bergerac* (1936), unexpectedly, in French, to Henri Cain's reduction of Edmond Rostand's play. One wonders why it took so long for Italian opera, always so eager to loot the stage for librettos, to get to Rostand's smash success of 1897, and in the event Alfano set the piece as if the artistic revolution of the 1910s and 1920s had never occurred, although the advance in technique from his rudimentary *Risurrezione* (*Resurrection,* 1904), a verismo debauch of Tolstoy's novel, is noteworthy. Somewhere along the way, Alfano had taken impressionism to heart, and he overthrew the clearly cut solos, duos, and Big Tunes for a continuous stream of Debussyan sound, translucently scored and pervious to text, yet bolstered by the resounding phraseology of Italian opera, a given south of the Alps since the days of Metastasio.

There would, eventually, be total modernism, from the likes of Luigi Dallapiccola and Goffredo Petrassi, but the old guard gave ground slowly. In 1930, when *Christophe Colomb, Von Heute auf Morgen,* and *Leben des Orest* thrust mixed-media, satire, and parody upon the operatic field, Italy enjoyed Franco Vittadini's *La Sagredo,* the libretto by Giuseppe Adami according to Puccinian strictures, centering on a hotblooded bluestocking in Napoleon's Venice who dallies for three ecstatic duets with a French lieutenant—but, alas, they must part. *La Sagredo* is a sort of nonviolent *Tosca,* with all three principals surviving the denouement, but a sort of *Tosca,* however neatly melodized, was no longer the thing. Vastly preferable, though equally forgotten today, was Lodovico Rocca's *Il Dibuc* (*The Dybbuk,* 1934), based on S. Ansky's famous play about a girl possessed by her dead lover, the whole defined by an "unbalanced" tonal aura and blessed with an unusual deviation in mode when the big duo unites the mad girl and her late lover in a scene at once tender and skin-crawling.

Cut off, unlike nondramatic music, from contemporary developments, Italian opera began to subside in reminiscence, lacking a vital younger generation to synthesize a renewed music drama. Surely, the neoclassical revival ought to have inspired a resurgence of Mediterranean polyphony, viewed in the facets of form, but such reorganization was allotted the tone poem and choral settings, not opera. Other nations were almost ready to wait for Godot; it was all the Italians could do simply to greet Pizzetti.

Something of an elder statesman by then, Pizzetti held to his kilter of legendary subjects fastened to a declamatory vocal line over choral and orchestral counterpoint, sadly never achieving a popular success. One after another, his operas landed and vanished: *Orséolo* (1935), *L'Oro* (*Gold*, 1947), *Vanna Lupa* (1949), *Ifigenia* (1951), *Cagliostro* (1953), *La Figlia di Iorio* (1954)—in sheer volume alone, Pizzetti must lead the ranks. How astute a musical dramatist he was came to be proved once again in his adaptation, no mere setting, of T. S. Eliot's *Murder in the Cathedral*, *L'Assassinio nella Cathedrale* (1958), for although Pizzetti did use Monsignor Alberto Castelli's translation of Eliot's play rather than a transformation of the material, he could not have succeeded so well had he simply cut the text and "composed" it in the manner of Debussy and Dukas' Maeterlinck operas. Too many modern operas are so created, out of unalloyed thespian greasepaint, and thus end up more as sung dramas than as musical theatre, overly conversational, *troppo staccato*. There is an essential difference between the dryness of atonalism, with its tone-clusters and its starting-and-stopping— its moments in time—and the accidental dryness of a text not meant to be sung. But Pizzetti, who never had any trouble in laying out his dialogue on a musical field, approached his task with an eye for operatic delineation, especially well sighted in the scene in which the four tempters visit Becket in Act I. Eliot, of course, had already provided Pizzetti with his chorus, the women of Canterbury, who irrupt between the second and third tempters while Becket soliloquizes, giving the opera an interior continuum that the spoken play inevitably must lack. Then, too, Pizzetti has the unexpected fourth tempter appear only as a shadow, for his bribe—an ambitious martyrdom—comes from within Becket himself, "to do the right deed for the wrong reason," and here again commentary from the chorus adds a second depth of characterization.

With so static an action, Pizzetti could not write to his usual great length, and unfortunately much of Eliot's poetic ironies are lost. The composer surrendered the chance to break into the profane in the scene in which the four knights justify their murder of Becket to the audience in coolly nonpoetic language; this Pizzetti reduced to a shred, and did not, as he might well have done, find a way to isolate this isolated moment from the rest of the pageant in musical terms. Worse luck, the sequence is usually cut in performance, emphasizing too much the "lives of the saints" aspect of the opera. True, Pizzetti always worked within a Christian worldview, one of ethical choices, rather than the blindfolded puppet Neverland of Godot that was spreading over the landscape in the late fifties, but he had no intention of turning Eliot's inquiry on hubris into a self-congratulatory miracle play.

Gian Francesco Malipiero, too, lent these years a continuity of output to match Pizzetti's in his own characteristic forms, as quaint and contained as Pizzetti's were numinous and immense. Heading into melodrama with his puppet figures and funny little tragedies—things of no moment, like life—Malipiero reprised the episodic structure of *Sette Canzoni* in *Torneo Notturno* (*Nocturnal Tournament*, 1930), seven vignettes of varying humor. This time, however, Malipiero does not range from one anonymous puppet to another; the seven "nocturnes" follow the paths of two principals, the Desperate Man and the Carefree Man, who enjoin the nightly contest from forest to tavern to prison, until at last the Desperate kills the Carefree and escapes from his dungeon to . . . liberty? No, no liberty for the Desperate, for he sees life too clearly to enjoy it. "It's not over," a backstage type announces, taking charge down at the footlights. To a snare-drum and tympani march, he tells the audience that a cor-

tege is going by behind the scenes: "It is life passing, waving
the flag of death. Listen." Now the orchestra assists with a march
that veers from bitonality to diatonicism and back again, herald-
ing the messy business of existence. But nothing happens on
stage. There is nothing to be seen, though the interlocutor
watches attentively. What? What? No answer: as the drums
fade, he vanishes.

Malipiero's axis turns on the intangibility of tangibles, the lie
of living—a fit subject for modern art, yet one that he composed
without a hint of atonal collapse. For Schoenberg, Berg, and their
disciples, the game is shudders and obsession; for Malipiero, the
game is gaming . . . there just aren't any winners. What, after
all, can a puppet win? How may a puppet choose? In expressionistic opera, beasts paw each other in misery, but Malipiero's evocative classicism doesn't kill the pathos with ostensive terror. His
sound, still tonal, still showing its fundamentals, still "clean," rejects the nightmare for the unsettling dream.

Acceding to the rise of national fervor in the thirties, Malipiero supplied his countrymen with a *Giulio Cesare* (1936) and
an *Antonio e Cleopatra* (1938), but he had not lost sight of his
own little world of nameless nobodies, and got Luigi Pirandello
to pen the text for *La Favola del Figlio Cambiato* (*The Tale of
the Changeling*, 1934), a ghoulish adventure in which a sturdy
peasant lad takes the place of the king's deformed son in order to
prove, as Pirandello often proved, that things are what they are
taken to be. "Nothing is true," says the imposter on his way to
his coronation, "and 'true' can be everything. Only God knows
for sure."

Right one is, if one thinks one is? Like the protagonist in Luigi
Dallapiccola's *Il Prigioniero* (*The Prisoner*, 1950), do we believe
that we are escaping our chains when in fact we're only being

led to the slaughter? Certainly Dallapiccola presents a convincing case in his atonal one-act, a landmark entry in the index of modern opera. The score, taken as an auditory sensation, is a perfect rack of agony and hallucination, but on the page it tends to serial discipline and parody: in one scene, the orchestra "improvises" three ricercares on motives of textual significance. So here is the classicist and his absolutism, but the libretto of *Il Prigioniero,* the composer's own, connects expressionism with the social documents of neoromantic theatre. As in Dallapiccola's *Canti di Prigionia,* wherein ancient prayers of Mary Stuart, Boethius, and Savonarola sing to music of modern provenance, *Il Prigioniero* connects the new with the old, in parody, in echo, in revival.

This, perhaps, is the ultimate of opera's modern voyage back into myth—neither to reinstil classic technique nor to resolve the mysteries of romanticism, but to synthesize the form of the first with the presentiments of the second—to resurrect the marvelous suspicions of the nineteenth century with the unmuddied vision of the eighteenth century. Yes, mannerism has intruded on the clarity in the person of the tone cluster and the gawking percussion of atonal drama, but this can be remedied.

Dallapiccola's atonalism is the useful, feeling music that renders drama intelligible, though he has thus rendered, in all, few dramas. His first opera, *Volo di Notte* (*Night Flight,* 1940), based on Antoine de Saint-Exupéry's *Vol de Nuit,* is set in an Argentinian airline office in the 1930s, during an aviator's disastrous attempt to pilot his craft in the dark. The score is keyed to mirror both the humdrum pokiness of the office and the questing determination of the men who suffer human defeat to win global victory, and the unseen, doomed aviator turned up as the central figure of *Il Prigioniero,* a more frankly serial work drawn from

Charles de Coster's *La Légende d'Eulenspiegel et de Lamme Goedzac* and from Comte Villiers de L'Isle-Adam's *La Torture par l'Espérance.* In this last hour in the life of a Flemish victim of the Spanish Inquisition, hope is the ultimate torture, hope kindled in the jailer's "brother," hope as the source of the prisoner's delusion of escaping to freedom, and hope chuckling at him when the illusion dissolves and the jailer reappears as a priest, again calling him brother but now bearing him to the stake.

Here's a familiar romantic notion: the human battle to overcome superior force forms, by the composer's own admission, the basis of Dallapiccola's oeuvre, whether in the quotidian industry of *Volo di Notte*'s office, in the expressionist prison of *Il Prigioniero,* or in the legendary excursion of Homer's hero in *Ulisse* (1967). In this evening-long work, however, the orchestra that sometimes refuses to grimace—and sometimes refuses to cease its infernal grimacing—loses the force of the two more concentrated short operas. Twelve scenes separated by the equivocating interludes one has come to expect of atonal opera, *Ulisse* is too dependent on the industrious percussion marked *grave* and *acuto,* on the gurgling celesta, on the sudden rushes and sudden hushes. Malipiero's more compliant musicality, while often played against the action, does carry the freight of dramatic movement, but Dallapicola's remorseless immobility weakens the picaresque *Ulisse,* especially since the spectator is expected to believe that the hero's experiences have enriched or depleted him, have made their journey along with him. Only one scene successfully qualifies inherited myth with contemporary psychological myth, when the enchantress Circe puts the finger on Ulysses, condemning him to live in Ithaca forever tormented by the lure of "the vast ocean."

Those who were content to work in less fully sublimated formats than those of Dallapiccola tasted of like disappointment, for though old-fashioned operas came along on a regular basis, none of them seemed to last out a minor flurry of regional interest. Terenzio Gargiulo's *Marie Antoinette* (1952), like Alfano's *Cyrano de Bergerac,* might well have belonged to the postverismo era but for the novelty of having the action narrated in flashback by a speaker, and Renzo Rossellini's *Il Vortice* (*The Vortex,* 1958) and *Uno Sguardo dal Ponte* (*A View from the Bridge,* 1961), from Arthur Miller's play, dripped blood from the rotting corpse of hoary urban naturalism. Luckily, Rossellini gave an ear to the impressionist school for *La Reine Morte* (*The Dead Queen,* 1973), a setting of Henri de Montherlant's dour play (the same plot that was to serve Thomas Pasatieri for *Ines de Castro*), amply suitable for Rossellini's reinvented Gallic ethos, being something of a *Pelléas et Mélisande* in less diaphanous exposition. Both *La Reine Morte* and the earlier *L'Annonce Faite à Marie* (*The Tidings Brought to Mary,* 1970), from Paul Claudel's play, marked something of a comeback for the composer and perhaps tokened a sign that verismo had at last retired its anachronistic parade on Italian stages.

Comedy, too, had trouble asserting itself, whether as pure frolic or in the more complicated mode of Malipiero. Dallapiccola's exact coeval and fellow modernist Goffredo Petrassi has won revivals but little real enthusiasm for *Il Cordovano* (*The Cordovan Screen,* 1949), his own rethreading of Cervantes, unhappily onto the spool of inappropriately drab methodism. Petrassi's limited musical scope sabotaged also *La Morte dell'-Aria* (*Death by Air,* 1950), a tragic-with-comic-interloping investigation of what impels the artistic quest, in the person of an inventor determined to leap off a tower to test his dubious parachute. On the other hand, an engaging divertissement based on Eugene Labiche's *Le Chapeau de Paille d'Italie,* Nino Rota's *Il Cappello di Paglia di Firenze (The Hat of Florentine Straw,* 1955) has traveled much of Europe with success. Rota, known more for his movie scores than his numerous operas, calculated an evening of jovial burlesque, hacking Rossini, spooking Bellini, mocking Verdi, and spoofing Puccini, yet never letting the spirit of Labiche's quite amusing original out of sight.

As in France, however, no one composer was able to maintain his continuity while pleasing the public, whose taste runs far behind that of the musical community. This, no doubt, is the void that Gian Carlo Menotti hoped to fill, and at the outset it seemed as if he might do so. His brief *Amelia al Ballo (Amelia Heading for the Ball,* 1937), a breathless banality about a woman desperate to get to a soirée who manipulates her husband and lover in vain and finally ends up, ball-bound, on the arm of the chief of police, did at least recapture the velocity of the old intermezzo, but Menotti really found his calling with a trio of shudderfests, *The Medium* (1946), *The Consul* (1950), and *The Saint of Bleecker Street* (1954).

Prolific and overrated because of his early success in America (where for some reason he is regarded as an American composer, despite his conspicuous Italian sound), Menotti writes his own librettos, and usually directs the premieres, adding to the popular impression that he is more a "theatre man" than a composer. His following among the critical establishment has never been large; a "poor-man's Puccini," they call him, though neither in subject matter nor in musical signature do they bear much comparison. More exactly, he is a poor-man's Menotti, and getting poorer lately with each new entry, for his operas appear to be built around a collection of stunts, the orchestra and voices fill-

ing in between the thrills like staged intermissions. Menotti's
work has largely originated in America, where the odious *Amahl
and the Night Visitors* (1951) is a fixture of the amateur masque,
and it is this, perhaps, that has subpoenaed a place for him in the
American annals, but in fact his music hasn't a shred of the national sound in it—*The Saint of Bleecker Street* might better be
done as *La Santa del Canale Grande*—and in general his output
is as ultramontain as the record of the French pop tune that
opens *The Consul* and the references to "granny" in his librettos.

All the same, one may at least admire *The Consul,* which for
all its willingness to tease with terror has the flair of barbaric
honesty about it. Arriving in the year of *Il Prigioniero,* it covered
similar territory in more accessible language and a coherent
storyline; here, too, hope is the ultimate torture. *The Consul's*
hopeful is Magda Sorel, trapped with her family in, it seems,
eastern Europe, and kept from fleeing the Communists by bu-
reaucratic indifference. Despite her pleas, and despite the fact
that her husband has been something of a freedom-fighter with
the underground, Magda never gets through to the consul, but
she keeps coming back, like others in the piece, hoping against
reason that she can appeal to the opaque consulate secretary and
her endless reel of red tape.

The Medium and *The Saint of Bleecker Street* have much
less to offer musically, though *The Medium's* little-show mobility should hold it in the repertory of the American pagliacci who
tour the high-school auditoriums of the hinterland on state
grants. Somehow, its Grand Guignol pseudo-occultism proves
compelling in the right hands, but while the music has dynamics,
it lacks presence, and since *Maria Golovin* (1958), Menotti's
operas have been lacking dynamics as well as presence. His comedy *The Last Savage* (1963) tried to poke fun at the consumer

civilization in the brittle style of *Amelia al Ballo* without recap-
turing its lightness, and the hollow *Help! Help! The Globolinks!*
(1968) sought to expose the bankrupt modishness of electronic
music by pitting tonal melody against invaders from outer space.
With melodies such as Menotti's to fight back with, earth is in
a lot of trouble.

A certain sympathy for the underdog haunted the outskirts of
most of Menotti's plots—and, of course, sparked *The Consul*—
but this sympathy became the core of Menotti's recent contributions to Third World liberalism, *The Most Important Man*
(1971) and *Tamu-Tamu* (1973), both of them fiascos of anti-
American drivel. Most recently, Menotti produced a "gentle
satire on a political theme," *The Hero* (1976), a measly charade
beefed up for length with cuts at middlebrow fashion, American
commercialism, and Watergate.

Sociopolitical shrillness has motivated also the *ragazzi terribili*
of the Italian avant garde. Hampered by either a lack of dramatic gift or the refusal to cultivate one, the revolutionaries seek
recourse in the "staged action," an excuse designed to relieve
them of the necessity to make sense. Luigi Nono's *Intolleranza
1960* (1961) is one such, a document of its time for musical per-
versity and leftist bilge. Another such, Luciano Berio's "staged
mass" *Passaggio* (*Passage*, 1963), offered its audience a soprano
undergoing unexplainable episodes in a pathetic life set to an
unlistenable score and harried by fellow cast members planted
in the audience whom she eventually dismisses. *Passaggio,* like it
or not, was at any rate a work that had to be composed; Berio's
so-called *Opera* (1970) didn't get that far. At its world premiere
at Santa Fe, its grunting nonentity was augmented by the fumbling collaboration of the Open Theatre of New York, a group
that specializes in improvised sketch-work.

Those who never venture beyond the frontier of early Strauss and late Puccini must learn that a tradition that stands still will only devour itself. "What is permitted"—Maeterlinck's Ariane, remember?—"teaches us nothing." Tradition, whether it be that of an art form, a national spirit as aired in art, or a mode, must be prompt in renewal, but destruction is not renewal. By tossing out both the drama and the music of purpose, today's avant garde manages to erase, for the moment of their evenings out, the two halves of music drama. They demolish the autotelic basis of music theatre, and what they are left with is ex, and ab, nihilo.

No, there must be a way to move onward and remain whole, to cut through the ice without ramming the ship; the experiments of the 1920s, in reviving the profane and larding it into the sacred, were sufficient unto that day, but what next? We can do without a plot, yet some sense of kinetic survey must survive, and, opera being a theatre for music, one way or another somebody is going to have to sing. That still leaves a lot of room to navigate, and as Malipiero illustrated in his trilogy on *The Children of Orpheus,* the word-music compartment can remain whole while the thematic power behind the text takes charge of renewal. For Malipiero, the puppets of Maeterlinck and Gothic Punch and Judy shows do not answer all the questions, so he retired the *commedia dell'arte*—the masks—and brought in the next genera-tion—human creatures, with no silly habits to fall back on and no merry dance with which to close their frolic. A child of Mali-piero would be Gino Negri, whose *Giovanni Sebastiano* (1970) deals with a Malipiero-type figure, a man who believes himself to be Bach, and who is successfully cured of his imagination only to die of shock (not unlike Wolf-Ferrari's Christopher Sly, if in wilder musical language). But the father himself is always wel-

come, and there was Malipiero himself, alive and hearty into
the 1970s. Reaching back to his old haunts, he wrote *Le Metamorfosi di Bonaventura* (*Bonaventura's Metamorphoses*, 1966)
and *Gli Eroi di Bonaventura* (*Bonaventura's Heroes*, 1969),
Bonaventura being a master of the grim-and-comic ceremonies
expected of Malipiero's puppet troupe, and the two evenings
apparently projected as the composer's farewell to art, for if *Le
Metamorfosi* gives us Bonaventura busy at his exercises in madness and merriment, *Gli Eroi* finishes with Bonaventura dying on
a pile of his own manuscripts after conjuring up ghosts from
Malipiero's earlier operas, earlier music and all.

In a way, then, Malipiero had organized his own tradition, and
with his own hand drawn its curtain. But at the same time, his
tradition is founded in the work of others—in the instinctive
framework of Italian canto, in the architectural diatonics of his
classical forebears, in the tenets of the comic intermezzo, in the
freakish tantivy of the romantics' identity seizures. Similarly,
others have drawn from Malipiero, or have come to life with him
as coeval partners, sharing in the evolution. Many of them teach
in the atonal college, where they are most comfortable expressing
the humid grotesqueries of their reckoning, but in whatever
style of music, their dramatic groundwork asks, insistently, who?
Ulysses? The prisoner? The changeling prince? The man who
would be Bach? Who?

Perky little codgers, Malipiero's puppets. So are we all.

3. GERMANY

More than the French and Italians, the Germanic artists suffer
modernism to waft over them, in both musical and structural

practice. No ultracontemporary stands out in the Germanic countries, as does Dallapiccola in Italy, for there is plenty of company. Godot, the ungod of modernism, the satirical tragedian who brays as he wounds, has disciples in this chapter, expressing the vacuity of life and the lie of free will—or the lie of predestination, depending—in operas at once literal and musical bombardments, and, too, the German audiences seem more inclined than the Latins to encounter difficult novelty. The remixing of the recipe that began with Wagner and continued with such as Debussy and Busoni continues still in the following pages, and we achieve at last the temporary consummation of tradition as current musicians and poets perceive it. The last years, the hopelessly up-to-date, are in sight.

Richard Strauss, for example, though he was to retreat from the neo-Wagnerian symphony in favor of the word quotient of *Intermezzo,* remained what he had so far been, closing his long life bound to tonality. Those operas of his canon already treated here are accepted as prime Strauss material, the others thought to constitute a curious decline, or, rather, a sudden and level step down, deficient in librettistic interest and melodic fecundity. Is it so? Neither Strauss nor Hofmannsthal added to the repertory with the diffuse and outlandish *Die Ägyptische Helena* (*The Egyptian Helen,* 1928), and *Arabella* (1933), Strauss' last Hofmannsthal collaboration, struck many as a pale replica of *Der Rosenkavalier.* The setting, again, is Vienna—1860 this time—with a cloddish air about the waltzes, and *Arabella* is, like the earlier work, a romantic comedy raised to some height by psychological subtleties in the principal soprano part, for the younger heroine shares with the Marschallin a serious view of herself and her surroundings. Unfortunately, though *Arabella* shows no less canny a musical narrative than its predecessor, it is less consistently attractive, with pages of texture creating ominous vacua between the "high points." And yet, so brilliant is Strauss, even in less than inspired melodies, that he is never the lowly craftsman, to be rebuked by neuter esteem and a few half-hearted musical examples. None of his later works falters as theatre, however seldom they astound as music (the heart does rather sink when the lovely finale of *Die Schweigsame Frau's* second act accidentally quotes from the "Chocolate" Waltz of *Der Rosenkavalier*). *Arabella* in particular, as it seems at last to be catching on outside German lands, draws one repeatedly closer to its two soprano sisters, the introspective, much-prized heroine and her generously emotioned, ignored little sibling, Zdenka, who spends most of the plot in pants because her parents can't afford to launch more than one debutante at Viennese prices.

No less a character study than was *Der Rosenkavalier, Arabella* exploits in smaller quality the synergic sweep of von Hofmannsthalian *gratia artis;* less of a work than its older sister, it is more of a piece, more consistent in its humors, more "correct" in tone. How is one to resist it, after all, however much one chides its passages that do rather go on? How to stand back and resist a comedy that diagrams its intrigues so artlessly and the great moments of which are so great as these?—the nick-of-time appearance of a wealthy aspirant to Arabella's hand, attracted to Vienna from his rustic estate by her photograph ("the pretty lady with the face," as his servant puts it); Arabella's first meeting with her fiancé that very night at her ball, and her last meeting with her three suitors, a last waltz for each; Zdenka's desperate gift of "the key to Arabella's room" to one of the three, who has been threatening suicide and whom she herself loves ("Silently she'll come to you," says Zdenka—oh ho!); that singular moment, after the ball, when Arabella's fiancé, who has overheard the key business, accuses her of perfidy and thereby so unhinges the comedy that Strauss withholds all music for a page or so, until poor little Zdenka comes running downstairs in her nightgown, "Zdenko" no more ("What a costume!" fumes her mother); and, at last, the famous "staircase" music when Arabella brings an untouched glass of water to her fiancé, a folkish token of her maidenhood, and he drinks it down and smashes it, the last of Arabella's suitors and her first and only man ("And you will remain as you are?" he asks her, and she: "I can be nothing else—take me as I am!"). And all that in one day, too.

Arabella was the end of von Hofmannsthal. The man who had bickered with, condescended to, and sustained Strauss in the most notable collaboration in all opera had died of a stroke just after the *Arabella* libretto was delivered. Worse luck, the fertile caverns of invention available in von Hofmannsthal's many prose works have not successfully been mined by other librettists; his contribution stops dead with this second Viennese comedy, and a royal humanist cedes the stage, heirless. Who else was able so skillfully to extract the humanity from the *gai primitif*? Next to his, other Zerbinettas are tawdry coquettes, other Ariadnes tedious with *sententiae*. He was a Chaucer of opera, and from here on, some capacity of cheer goes out of the narrative.

All for comedy now, having enjoyed his bout with the Word in *Ariadne auf Naxos* and *Intermezzo*, Strauss tried another dip into the profane with a parody of the true *opera buffa*, frantic, insolent, and crackling with wordplay and alarums. This was *Die Schweigsame Frau* (*The Silent Wife*, 1935), with a libretto by Stefan Zweig culled from Ben Jonson's *Epicoene; or, The Silent Woman*, said woman being the matrimonial objective of a man who can't abide noise of any kind. Given the premises of *opera buffa,* he is doomed to suffer noises of all kinds, naturally, and Zweig overweighs the poor man's burden with a turn out of *Don Pasquale:* once married, the silent wife becomes a virago until her husband yields her to his young nephew, as the logic of comedy demands. (The logic of Jonson's original resulted in transsexual skulduggery, but opera wasn't quite ready for that; anyway, Straussian opera was not.)

Here the composer caught up with neoclassicism, though entirely through his own métier and, as ever, at the feet of Mozart and Wagner—the Austrian, be it said, dominating the lecture. *Die Schweigsame Frau* entails the structure of closed-ended opera while opening the ends: a potpourri overture in one breathless movement leads off, arias and ensembles trespass the way they used to and, to create the intrigue, there is even a barber on hand—if not of Seville, at least cut of the cloth. Parody not only decrees the overall form but lavishes pastiche within the framework, in two extracts from The Fitzwilliam Virginal Book and in drop-ins from *L'Incoronazione di Poppea* and Giovanni Legrenzi's *Eteocle e Polinice.* Pastiche as well colors the complexion of Strauss' own music, as when he commences the scheming finale of Act I with the theme that had opened the overture, now in an insinuating minor key with the woodwinds feeding the conspiracy and the players interrupting the Barber's plan, jumping in with the baggage of Paisiello and Piccinni until the finale proper gets going in a rousing G Major of stretta and counter-stretta. The plan of the recitative is Wagnerian rather than secco, yes, but the method of musical organization invariably points back to the happy days when myth was no more than graceful diversion and the Hofmannsthals philosophized in prose. Never again was Strauss to confront the avatars of subtextual depth on their own terms—the impenetrable Egyptian

Helens, the shadowless women and their shadowy initiations. He obviously enjoyed the sunny clarity of Zweig's farce, propelling the dominant right to the tonic as if tonality were going out of style. And, of course, it was.

But not *bei* Strauss. Tonal to the end, and even waxing slightly less polyphonic, he unhappily lost the partnership of the excellent Zweig, a non-Aryan, and turned to the acceptable Josef Gregor for *Friedenstag* (*Day of Peace*, 1938), an appalling one-act set in the garrison of a city losing a siege during the Thirty Years' War. Each premiere of a Strauss opera was a national event, and *Friedenstag* is something of a national piece (though its antiwar message couldn't have been less nationally apropos at the time), not least in the little barcarolle sung by an Italian emissary, expressing an outsider's *fatigue du nord,* and in a loud C Major finale in homage to *Fidelio* that sports one of the most ungainly top soprano lines in all opera. Much easier on the ear was the "bucolic tragedy," *Daphne* (1938), metamorphosis and all, a sign of the "new" classical Strauss, departing the athletic psychopathy of *Elektra* and *Die Ägyptische Helena* to build on the humanistic warmth of *Ariadne auf Naxos*. Among the ideas that Strauss and Gregor projected were a Nausicaa, an Amphitryon, and a Semiramis; eventually they arrived at Danae, the girl whom Zeus ravished in the form of a shower of gold. Adding in Midas to complete a triangle, Gregor gave Strauss *Die Liebe der Danae* (*The Love of Danae*) as his Greek offering. Finished in 1940, rehearsed in 1944 at Salzburg but canceled for Hitler's Total War, Danae saw light at last in 1952, as the final premiere of a major Strauss opera and the closing of the era of classical chastity. One year before, Stravinsky had redefined neoclassical parody in the uninhibited contours of *The Rake's Progress,* and *Penelope, Pallas Athene Weint,* and *The*

Bassarids, very modern antiquity, were soon to come, so the gentle *Danae,* gentle in song and story, really marked the ultimate of Greece refined.

Die Liebe der Danae joined *Daphne* in simplicity. Strauss had always said that the gargantuan *Elektra* could be word-clear if only conductors would observe his markings, but even so the Greece of the late Strauss operas is far more restrained, striving for elegance and lyricism, and the immobile pervertedness of Hofmannsthal's Elektra and Helena—and, less noticeably, Zerbinetta—now gave way to the interior growth of Hofmannsthal's Ariadne. Gregor's Daphne becomes one with the green world, and his Danae and Midas, afflicted with gold-love, develop as people by discovering human love. Heard from as infrequently as *Die Ägyptische Helena* and *Friedenstag, Die Liebe der Danae* deserves better treatment, not only for its eloquent romance and its powerful characterization of Jupiter (Zeus, more properly), but for its complement of satirical jollity in a quartet of the god's cast-off flames—Semele, Europa, Alkmene, and Leda—all of whom are ready to leave their husbands for another round with the Olympian Casanova.

Now comes the dessert, Strauss' opera about opera, *Capriccio* (1942). Collaborating with Clemens Krauss, a Straussian conductor of credentials almost rivaling those of the composer himself, Strauss penned a libretto set in the salon of a chateau near Paris at the time of Gluck's revolution in music drama, just when the music was asserting its place in the entity and subverting tradition. A timely piece, *Capriccio* erupted with classical reminiscence just before a serial catastrophe was threatening to force tradition to commit seppuku with the short-handled blade of unmelody, and offered a sort of congeries of Strauss' personal tradition. The premise: a group of artists gather at the house of a count and his sister, where matters of art and art's future are discussed. A poet, a composer, a theatre director, and an actress represent the arts, and while the count prefers the spoken drama —and the actress—to music drama, the countess cannot decide between music and poetry, and solves her quandary by playing the muse to bring the poet and the composer from friendly enmity to mutual respect and collaboration. With talk of reform opera, buffa, and seria in the air, the two men are convinced to write an opera themselves and the count, of all people, suggests an unlikely project: let them write an opera about the events of that very day.

Capriccio, in other words, is about how *Capriccio* was written, not just by Strauss and Krauss, but by the concatenation of aesthetics and eras, by all the Strausses and Krausses that ever were, by all the events leading up to the comedy—the *querelle des bouffons,* the Gluckian orchestra, the romantic leitmotif, the development of through-composition and thematic psychology, the revitalization of parlando and fantasy, the reformation of neoclassicism. Capping, it would seem, not only his career but that of all opera as well, Strauss made his farewell to opera with a holiday of pastiche and quotation, including recollections of Daphne, Ariadne, and the composer's theme from *Ariadne's* prologue, plus a morsel of *Der Rosenkavalier* tucked into the Countess' closing monologue. Pascal and Ronsard are cited, as is Gluck's *Iphigénie en Aulide,* and while most of the text appears to be a battle between le parole and la musica, a modern-day *querelle des bouffons,* the question is really settled in favor of neither by the director, La Roche, who hectors the poet and composer in a lengthy speech on the *Gesamtkunstwerk* circa 1775, on "the eternal laws of the theatre," knowing neither literary opera nor symphonic opera, only great opera. "Piously I cherish

the old," cries La Roche, "patiently awaiting the fruitful new,
watching for the masterworks of our age!" He is Strauss' spokes-
man, this man of the stage, not the composer, Flamand, nor yet
the poet, Olivier; opening, as Olivier puts it, "the cupboard of his
rich experience," La Roche has seen it done every which way,
scrambled, fried, poached, and boiled, and his recipe is Strauss'.
Why, even his good-natured antiintellectualism reminds one of
the Strauss who more than once set Hofmannsthal's stage directions along with the poetry:

> *Seht hin auf die niederen Possen,*
> *An denen unsere Hauptstadt sich ergötzt.*
> *Die Grimasse ist ihr Wahrzeichen—*
> *Die Parodie ihr Element—*
> *Ihr Inhalt sittenlose Frechheit!*
> *Tölpisch und rüde sind ihre Spässe!*
> *Die Masken zwar sind gefallen,*
> *Doch Fratzen seht ihr statt Menschenantlitze!*

> Look at the low farces
> In which our capital delights.
> The grimace is their token—
> Parody their style—
> Their content feckless insolence!
> Loutish and coarse are their jokes!
> Yes, the masks are down,
> But one sees caricatures instead of human faces!

The masks, of course, are those of Jonny the bandleader, of the
headstrong Nose, of Doktor Faust and Mack the Knife . . . of
Wozzeck and Lulu, even. Not these gargoyles for La Roche:

> *Ich will meine Bühne mit Menschen bevölkern!*
> *Mit Menschen, die uns gleichen,*
> *Die unsere Sprache sprechen!*

Ihre Leiden sollen uns rühren
Und ihre Freuden uns tief bewegen!
Auf! Erhebt euch und schafft mir die Werke, die ich suche!

I want to people my stage with human beings!
With humans like us,
Who speak our language!
Their sorrows should move us
And their joys thrill us to the utmost!
Up! Arise and create for me the works that I seek!

Capriccio's libretto is a trifle mannered and its music, again, not exclusively scintillating, but the piece is engaging enough for all that, and a suitable farewell given Strauss' advance into the cool pasture of Mozartean fundamentals. And in a way, the inside joke of the plot-about-a-plot, the mirror within the mirror, is one method of meeting the moderns on one's own doorstep. Strauss actually concocts his own extremely undistinguished exit, when La Roche, departing with Flamand and Olivier, coaches them in the theatrical secrets their operas must unlock. "Above all," he solemnizes, "arrange for good exits in my role! You know, the telling exit . . . the last impression of a character. . . ." And on that note, squandered in the wings, he vanishes.

Strauss was the only romantic to dominate the German opera scene in the thirties and forties, despite a buzzing hive of romantic bees, most of them closer to Korngold than to Strauss in sound and subject matter. The sigh, the desire, the ghostly freemasonry of the *Schauerromantik* were still the rage, but Strauss' most adept fellows had gone over to species of modernism. Arnold Schoenberg and Alban Berg stuck with atonalism (Berg even adopted serialism for his second opera), which also had begun to entrance Ernst Křenek, and Carl Orff was about to trade in his folkish harp for the clamorous brouhaha of the Greek percussion section. Only Paul Hindemith resisted tonal overthrow, filtering the formalism of *Cardillac* with the equally neoclassical but less angular *Mathis der Maler* (*Mathis the Painter*, 1938) and *Die Harmonie der Welt* (*The Harmony of the World*, 1957). Choosing as his protagonists a painter and an astronomer and for his time-scheme the late Renaissance, Hindemith made this pair his *Palestrina*, his *Doktor Faust*, writing his own librettos in troubled, megapolitical times when the role of the artist in society was made the object of some heavy analysis by artists and politicians alike. Hindemith found for art, not politics. His painter Mathis, modeled on the real-life Mathis Grünewald, is caught up in the upheaval of the sixteenth-century Peasants' War, when it appears that action speaks more loudly than art until Mathis envisions himself as Anthony undergoing the temptations, resisting, and finally receiving the advice of Paul the Apostle to seek his own action in the immortality of art, committing his analysis to canvas. In a passage as enthralling as that which Pfitzner devised for the vision in *Palestrina* (though much less grandiose), "Paul" and "Anthony" raise their voices in alleluias on modal triads in a broad $\frac{3}{2}$—although one might ask how Hindemith could possibly have thought to reconcile his crescents of piety with the ogreish horror of the real Grünewald's St. Anthony triptych.

"Cultural Bolshevism" the Nazis dubbed it, stupidly lumping Hindemith with Křenek and Weill, and the creative Mathis with the destructive Jonny and Mack the Knife. Not as easy to enjoy as *Die Meistersinger,* the slow-moving *Mathis der Maler* is no less dedicated to the glory of national art, even if Wagner celebrated the musical revolution and Hindemith preferred the old order. *Mathis* has too often been labeled, first by gangster-politicians, then by gangster-critics, who deem it "undramatic." In fact, the libretto *is* undramatic, but the music most definitely is not, as a well-prepared performance of the piece always proves. Unfortunately, the lie is perpetuated by *Die Harmonie der Welt,* for this sagging colossus lacks the momentum of the earlier score, and does indeed show itself sessile on stage. Its star-gazing hero, Johannes Kepler, is, like Mathis, based on a human predecessor, its title derived from the real Kepler's dissertation, *Harmonices Mundi,* and the opera as an entity does bear an uncomfortable likeness to a medieval treatise, coming off in its fourteen friezelike episodes as metaphysical argument rather than theatre. Actually, Kepler/Hindemith's postulate, that music-order is world-order, is a ripe subject for sacred opera, and the composer did plot out a dazzling commencement and finale in the form of a toccata and passacaglia, respectively in e minor and E Major; alas, in between lies stationary gesture.

The more outrageous practitioners of neoclassicism, Kurt Weill and Ernst Křenek, changed tack suddenly in the thirties. After *Aufstieg und Fall der Stadt Mahagonny,* Weill and Brecht could go no farther in that direction; they pulled back for a cantata on Charles Lindbergh's flight from New York to Paris, *Der Lindberghflug,* and a one-act for school use, *Der Jasager* (*The Yes-Sayer*), both in 1930 and neither overtly vicious. Not long after, Weill had left Germany for the free world, where he applied Brechtian *spielsingen* to Paul Green's text for *Johnny Johnson* (1936), and then, abruptly, switched styles to work within the traditions of the American musical, instituting a few traditions of his own from work to work.

Křenek, too, abandoned his shirty *Zeitopern,* specifically for psychological pageant in *Karl V* (1938), a "stage work with music" partly serial, partly atonal, partly spoken, and partly in C Major—for Křenek, in his route through style, never declared himself for any one camp. *Karl V*, unlike Křenek's *Leben des Orest*, did not mix modern tomfoolery in its period narration (though the composer couldn't resist the saxophone even here), but then it didn't have to, the period of *Karl V* being possibly the most grotesque in Europe's history. Křenek's libretto takes in Martin Luther, Moritz of Saxony, Pope Clement VII, Francis I of France, Ferdinand and Isabella, Pizarro, clerics, sailors, a chorus of the dead, and even the Voice of God, tenors in unison who call the abdicating Carlos to judgment in the first scene ("Too soon retired, Carlos—no more Emperor, no more king: man before My highest throne!"). Carlos dies in the final measures of the long opera, but not before the Reformation in all its conniving has been held up as a concept of art-life, and not before the complicated Carlos has been conceived of along with it. "With this man an epoch dies," says one of the characters at the end; with this opera a new epoch is continued, and the final chord, played by the strings, registers all twelve tones of the scale.

Křenek's unorthodox use of dodecaphonic organization remained his sound right through his later career and, still writing his own librettos, he tried to locate a level of audience involvement that would impress poetically yet not threaten one's sense of humor. Such an air was to have lifted *Pallas Athene Weint* (*Pallas Athena Weeps*, 1955), but Křenek's fierce defense of peace and liberty set in the Athens of Socrates weeps too much itself in the atonal fragments that by then had lost their novelty and, at least in this case, their dramatic effectiveness. After an American interlude, *The Bell Tower* (1957), adapted from Herman Melville, Křenek recaptured the absurd social commentary

of his *enfance terrible* in *Das Kommt Davon, oder Wenn Sardakai Auf Reisen Geht* (*It All Comes Out; or, When Sardakai Goes On Her Tour*, 1970), a satirical potshot at contemporary life, including women's liberation, leftist angst, psychoanalysis, and media blitz. Again, however, the music lacks the panache that Křenek wore when the vulgar *Jonny Spielt Auf* put him on the world stage.

Strauss, the romantic turned to a limited classicism; Hindemith, the sacramental formalist; Weill and Křenek, the gadflies who defected, one to Broadway and the other to nondramatic atonalism: these were those principals of early twentieth-century opera who survived into the chaos of postneoclassical confusion, when German music and theatre craft took the Western opera leadership away from France and Italy. And where was this leadership to take us? Busoni and Pfitzner, who in different ways nurtured inheritances of tradition, had run their courses, and Schoenberg and Berg left only one new work, each unfinished, at their deaths. Now must come the next generation in German opera, and if much of it had obviously listened carefully when *Erwartung* and *Wozzeck* were playing, they had all heard Stravinsky, and Hindemith and Busoni as well. Where will they take us? They must take us onward, past and forward. That is what leadership is.

But before the newcomers had announced themselves, in 1937, came the premiere of Berg's *Lulu,* and twenty years later, after they had arrived, came the stage premiere of Schoenberg's *Moses und Aron,* and in a way these two had more to say about opera's future than the fresh titles that followed them, and about the tailoring of preceptive fantasy and realism for the lyric stage. And in another way, these two prepared final statements on the wedding of music and word so distinctively as to have both bewitched and bewildered operagoers ever since. *Lulu* is intimate and restless, a reasonably contemporary drama about a woman; *Moses und Aron* is huge and measured, theatre of ideas invested with Judeo-Christian myth. They stand, together, as neither the latest nor the last events of their kind, bestriding the emergence of younger talents. To some they were prototypes; to others, gravestones. And everyone has been consumed with the need to decide whether they are to be imitated, analyzed, or shunned.

Wozzeck had not been scored for serial technique (or even for a genuine atonal sound), but in his second work Berg joined the twelve-tone establishment, retaining the forms-within-forms of *Wozzeck* in a more fluid structure and allowing for the motivic portraiture that for all its overuse has never lost its usefulness. For *Lulu's* libretto he again consolidated avant-garde drama of the past, Frank Wedekind's *Erdgeist* and *Die Büchse der Pandora*, both of which center upon the rise and fall of a fatal woman, a Lilith fit for primitive moon-worship, lovely, implacable, and devouring. A coloratura soprano in the opera, she becomes almost human, and despite the degradation she inspires in those who adore her, by the end of the evening she has added the spectator to her conquests; music makes her fascinating in the dimension unseen but sensed, augmenting what we can see of her on stage. The libretto tells us she is fascinating; the music can prove it.

Schoenberg, too, wrote his own words, and his hero is Moses, leader of a people rather than a self-oriented individual, and self-willing rather than instinctive, drafted for history in the first scene by an undulating chorus representing the burning bush. Lulu, on the other hand, inaugurates her story flirting with a painter and seeing her husband die of a stroke when he surprises them together. It continues so throughout in the two works, as

Moses goes on to lead his people out of bondage and into the desert and a crowd scene of astounding proportion re-creates the orgy around the golden calf and Moses' smashing of the tablets, whereas Lulu leads no one, not even herself, but simply goes on ruining lives on a sparsely populated stage until she ends as a prostitute in London, where Jack the Ripper closes her career.

These two very different works were composed at the same time, circa 1928-34, and grew out of the same revolution of musical dramatization, the linear narrative revitalized by nonharmonic through-composition and song-speech—that is, by post-Wagnerian opera with the developmental texture and double-bar lines removed. The leadership of the newcomers would not have to take opera all that far along after all, for both Schoenberg and Berg had already taken it there. Serialism may look like a stunt, but serialism in *Lulu* and *Moses und Aron* is unerring procedure, and ultimately it removes the final barriers that defeated the cyclic unity of Wagnerian opera, that made of it a sacred unity but a unity in default of verbal completion. Wagner himself proved this in *Parsifal*, which moves into modernism but half-armed: the speech is still song, and thus works neither as song nor as speech; the closed stops are gone, but gaps, not continuity, replace them; and even the myth fails, for its redemption is only magic, and its ethic legislated rather than organic. Transformation, however, if not inevitable, is essential, and moments of *Parsifal*, such as the first bars of Act III, abstract the attainment that Schoenberg was then to theorize and practice in music that "never repeats" and yet is always more of itself, like a sheet of glass.

For theatre, serialism can easily fail, but then so can tonality. Method in *Lulu* and *Moses und Aron* is theoretically "objective" and noncommunicative, but in practice it is wholly theatrical communication, and contains the ambiguity that has joined old irony in layering the meat below the tissue of drama in the modern age. Lulu is Eve (called so by one of the characters, in fact, though introduced to the public in a prologue as "our serpent") and human, all too human—avatar and individual, not by turns but simultaneously. Was there any tonality left in the twentieth century to accommodate her, even in Skryabin's harmony and Stravinsky's metre? When her husband is pounding at the door in the first scene, about to catch her with the painter, Lulu panics. "He'll beat me to death!" she screams—but moments later he is dead, and when the painter leaves to fetch a doctor, she eyes the corpse. "Pussy," she calls now. "Oh, he's only fooling." She circles the body, saying, "He sees my feet, and watches every move I make. He's got me in view all the time." She stretches out her toe and jostles him. "Pussy!" But her husband is dead after all. "What should I start now?" she asks.

Moses is not so mercurial, his music less capricious, but the same musical fluency guides the minutes. Unlike the sequences of the romantic scores, this music arcs in one spasm, and no separate tread can be detected. Yet there are arias, and scenes. There is ballet and effect. Where *Lulu* drops the eternal White Goddess into a quasisurrealistic naturalism, *Moses und Aron* inhabits the dour frontier of fantastic antiquity in a well-marked maze of philosophy and psychology. Moses, as he tells the burning bush in scene one, "can think but not speak," and while Aron is to speak for him, Aron has a will of his own and speaks as perfectly for gold and Aron as he does for Moses and God. "Oh word, you word that I lack!" cries Moses, trapped by the easy images of Aron's surface communication, images that obscure the one image of God, the unproduceable.

Schoenberg, they like to say, was also trapped by his thralldom to his one image, to a word that he, too, lacked. They compare the composer to his hero, make dodecaphony the unproduceable God and Schoenberg the Moses who can think but not speak, who didn't want to lead and didn't do it well. But his music does communicate. It can be produced. Yet it does without the interior proportions that Western music grew up on—the chord, the melodic period, the repetition, the movement. In the end, Schoenberg found a way to unite Moses and Aron in the very premise of twelve-tone opera, for in the unlimited manifestations of the note row, he combines the infinite capacity of music with the bull's-eye immediacy of speech—yes, the sacred with the profane. This, at last, is what had become of the story-ceremonial, and though *Moses und Aron* tells in language far removed from that of *Kityezh* and *Doktor Faust,* its intent is the same.

Interestingly, for the autobiographical theory of Schoenberg-as-Moses, he for some reason could not bring himself to set his brief third act, and ended composition at Moses' "Oh word, you word that I lack!," which does rather second the assertion that Schoenberg saw himself as Moses. Berg, too, left the third act of *Lulu* not quite done, the short score clearly laid out but not fully orchestrated—although Berg ceased operation through death rather than any mystical process of artistic *versteinen* that one might care to allege. *Lulu* does show the traces of self-dramatization, nonetheless, when Alwa, the son of one of Lulu's victims and himself a victim-to-be, says of her, "Surely an interesting opera could be written about her," and goes on to project a few scenes of a "Lulu" libretto—each one right out of Berg's own plan as we have so far seen it unfold. Alwa even predicts, correctly, the next tragedy about to befall Lulu's admiring circle. If Berg's Act III is ever presented as he designed it, Lulu's victims will get their licks in, for the four men destroyed by her in the first half of the work are to reappear as her four clients in Act III, one of them to commit the climactic gesture as Jack the Ripper.

With these two blockbusters haunting their progress, the younger composers had much to live up to—imitate, analyze, or shun—whether their allegiance held to romanticism, neoclassicism, atonalism, or the ultramontain influences of Debussy and Stravinsky, and whether they preferred music or the Word in whatever ratio. Among the first to surface were Werner Egk and Carl Orff, both in a burst of ultratonal music seeking a diatonic clearness which Orff was later to lose by straying altogether too effectively into the percussion section. In this technological age it is fitting for composers to conceive works for the airwaves, and Egk's first opera *Die Zaubergeige* (*The Magic Violin,* 1935) set a fashion that has interested men as disparate as Douglas Moore, Pizzetti, Menotti, Britten, and Henze, but which, even with television at its disposal, has made no real contribution to opera other than in the matter of mass exposure. After this Humperdinckian piece, Egk adapted Ibsen in *Peer Gynt* (1938), and this, too, proved a fairytale where the Norwegian original is meant as pungent satire. *Peer Gynt* would make a corker of an opera, and one day may do so; Egk's version is too contrivedly operatic.

Audiences enjoy Egk's folksy, amiable ambles through formula, but his *Columbus* (1942) was officially deemed an inferior likeness of Milhaud's *Christophe Colomb,* which no one likes anyway, and his *Kirke* (1948), based on a prankish Calderón original, was withdrawn early on. His own librettist, eager for adaptations, Egk delivered Yeats and Gogol up to the musical stage in his two most resourceful works, *Irische Legende* (*Irish*

Legend, 1955) and *Der Revisor* (*The Inspector,* 1957), both of which are semiregular items on the German agendas. Still, the cognoscenti deem him antediluvian, and went out of their way to disdain his 1966 revision of *Kirke, 17 Tage und 4 Minuten* (*17 Days and 4 Minutes*), the title referring to the time that elapses on Circe's island before her enchantment holds all visitors under her sway forever. In keeping with the frolicsome air of the text, Egk bedeviled his second version with a pride of pastiche numbers, not excluding a touch of very old-fashioned rock.

Egk's Bavarian countryman, Carl Orff, has already been mentioned for his two storybook parables, *Der Mond* and *Die Kluge,* but here he assumes a more messianic position as the teacher of a primitive music drama, looking back to the monodic recitative of the Camerata and even past that early juncture to the Greek festival, in three titanic and largely unlistenable works, *Antigonae* (1949), *Oedipus der Tyran* (1959), and *Prometheus* (1968). Securing no real libretto such as we today understand the word, but simply setting Hölderlin's translations of Sophocles and Aeschylus, Orff tarried at the flat edge of the earth in *Antigonae* with relentless ostinato patterns out of some primeval mustering, and then dropped off into space with the virtually musicless *Oedipus* and *Prometheus,* the singers' lines hurled out in feral plainchant and the audience, generally, hurling themselves into and up the aisles long before the curtain. It's odd that, in a time when opera has rediscovered myth for psychological spell-casting and hard-nosed commentary, Orff should have followed his early folk pieces with such flat and unadorned presentation. In the matter of words, his method is no different from that of a number of modernists, but the scores lack any formal tension once the novelty has worn off. Perhaps Orff intended to show naked man

in naked music, and it is arguable that his Greek works return
the plant of opera to its original Greek cultivation as chanted
musical theatre, but such allowances do not defend the minimal
musico-dramatic operation of *Oedipus* and *Prometheus,* if not
Antigonae, and one notes that in the more recent *De Temporum
Fine Comoedia (The Play of the End of the World,* 1973), Orff
fell even more deeply into the void by setting a text in Greek
and Latin.

Of those who seized the haft of late-romantic tradition and
have attempted to cut their way through to formal security,
Gottfried von Einem is perhaps the best known, mainly because
of a good instinct for theatrical ploy and an essentially conservative idiom highlighted by recollections of the jazz age, both of
which give his works the air of being the current *Zeitoper.* Like
Hindemith, Křenek, and Weill in the twenties, von Einem veers
to the sociopolitical, though in unthreateningly vague terms.
Dantons Tod (Danton's Death, 1947), *Der Prozess (The Trial,*
1953), and *Der Besuch der Alten Dame (The Old Lady's Visit,*
1971) all deal in one way or another with the corruption of the
System, all come translated from other media, and all have scored
some success in Europe. Egk and Orff had announced themselves
before the war, along with such others as Hermann Reutter and
Rudolf Wagner-Régeny; von Einem, however, signaled the arrival of a postwar generation, and there had been much worry as
to what the psychological bender and hangover of the upheaval
would cost artistically. Much, therefore, hung on the Salzburg
premiere of *Dantons Tod,* and while its source as Georg Büchner's most brilliant—certainly most exciting—play led many to
compare the opera unfavorably with *Wozzeck,* many others were
glad it had come out as well as it did.

Büchner's view of the French Revolution seen from certain

pairs of eyes made no great impression as purely thematic opera, for the hectic love-death of the characters doesn't sound quite as committed as it might have done, but *Der Prozess*, Kafka's tale of an innocuous citizen put on a sort of trial for no specified crime, brought out the caginess of ghoulish satire, if without the ultimate in Weill-Brechtian irony. Von Einem's teacher, Boris Blacher, himself a composer, and Heinz von Cramer settled Kafka into nine scenes, each worked by von Einem in a different mood, rhythm, and tonality, though that last word is becoming basically inappropriate. For so episodic a libretto, one with presence but little suspense, von Einem's syncopated freakishness is wealthy with the exactness of atmosphere that *Dantons Tod* lacks, especially in the opening, when a stranger strides into the protagonist's bedroom and a few lines of metred speech turn into one-note lines of recitative as the orchestra wanders into an accompaniment bit by bit, reality turning into opera by the measure.

Given the scope of subject matter in German opera, one expects the great themes of the age to ignite its librettos just as Busoni had to present his personal Faust and Pfitzner his personal curator of the *Zeitgeist, Palestrina.* It was inevitable that someone attend to Kafka's nightmare in the afterbirth of the modern war-nightmare, and, in the same vein, von Einem tackled Friedrich Dürrenmatt's play *Der Besuch der Alten Dame*, the lady of the title being the world's richest woman and her visit a gala return to her seedy little Swiss hometown, where she offers an incredible fortune to the people provided they murder the man who seduced and abandoned her years before. Dürrenmatt's drama is a modern classic, the play Alfred Lunt and Lynn Fontanne chose to make their farewell in, and a balloon-scratching, masterfully beastly work it is, as the good townsfolk,

lured by wealth, gradually get used to the idea of their own du-
plicity and accept what is. This was von Einem's major challenge
to date, for any number of composers can make an adequate
opera out of a feverish Kafka piece—many have tried—but a play
as fine as Dürrenmatt's does not necessarily make a dream libretto. By leavening the "talking scenes" with cantilena, von
Einem brought it off. His revolutionary mob in *Dantons Tod*
seemed too much the operatic villagers, ready to help a guillotine
scene along with a Carmagnole, but this modern Swiss mob is,
rightly, more perfunctory and thus all the more repulsive. At the
end, after dispatching their sacrifice, they break into a dance of
triumph, punctuated by ecstatic imbecile sounds, no words—
no words necessary.

Though operas along the lines of Bécaud's *Opéra d'Aran* and
Rossellini's *Uno Sguardo dal Ponte* proliferated in the 1940s and
1950s, as they had done in the 1890s, another overturn in form
and subject matter was clearly on the way, and opera was now to
catch up with the modern theatre in the mixed genre of poetic
individuality cued for the helter-skelter of comedy, the mode of
Samuel Beckett in *Waiting for Godot*. We have seen it coming,
stealing up, often so frontally, so full of itself and unbashful,
that one knew better than to put up a fight. It came; it conquered; but it does not see, for it knows how much of the old
theatre questions cannot be served by the old theatre answers. It
knows that there are no answers and thus makes no attempt to
see, to comprehend. It presents. Here is von Einem's Old Lady,
promising her billions to her former fellow citizens for the pun-
ishment of her former lover with a sense of irony and no doubt
in the least, living for life's facts and asking no questions of it.
Here is Josef K., the center of *Der Prozess*, innocent and yet
executed—not wrongly, for no crime has been committed, but

summarily, matter-of-factly, and with no answer to his question. Here is Berg's Lulu, a fact if there ever was one, ending her fall in a retrograde transformation of her rise, her victims returned as her assailants, yet it never occurs to her to ask about that. Why should she? Who would answer—one of the other puppets?

As late as *Louise* in 1900, nothing in opera had changed all that much despite the shocking working-class naturalism, but as early as *Pelléas et Mélisande* in 1902 opera had changed drastically, moving back into fantasy, and then into legend, and now into the new myths—the old ones, really, except taken out of the heroic context and deprived of the grandeur of tackling fate, of being a king, or even of knowing one. Some of today's heroes don't even have names, like Dallapiccola's Prisoner, and even when they do, like Dallapiccola's Ulysses, they are not the titans we associate with the titles. Here, while the current generation in German opera unveils itself, is exactly where to stop and realize that patterns set up in the early twentieth century have fulfilled themselves at last, and that both subject matter and musical mode—which, put together, equal form—have in combination linked the quest of romanticism to the finality of neoclassicism, the despair of one with the logic of the other . . . the logic of *logos:* the Word. The stages of the St. Germain and St. Laurent fairgrounds have met the hero after circling around him for years; the glory meets the grimace, and fancy destinies give way to the wilderness here and now, to the restless irony of double-genre and self-contained intrusion. What music expressed once, music still expresses, but the wonder is gone, and technique must take over for passion. Instead of action, the point is communication—an instant rather than an adventure, a fact rather than a dream. The dreams today are anonymous, and the facts are those of the clowns.

The stories are familiar to us by now, stories of scrambled identities that never come unscrambled or of nobility squashed with *Spass*. In Rudolf Weishappel's *König Nicolo* (*King Nicolo*, 1972), from Frank Wedekind's play, a deposed king wangles his way back into court by turning jester, and proves so convincing in the role that no one believes him when he reveals who he really is. In Rolf Liebermann's *Penelope* (1954), Odysseus, Achilles, Telemachus, and the famous waiting wife are guyed with derision: Penelope has married during her husband's absence, and he dies of a heart attack from the twenty-year exertion of body and mind. It is the pointed, now-burlesque, now-elegy pandemonium of opera's contemporary musical fittings that presides over the event, and even the less directly ridiculed characters in the more traditional librettos must contend with the shadowy cackle of atonalism more frequently than not.

This is not to say that modern German opera is all innuendo and puppet play. But in any given moment in art, there are those who pledge themselves to orthodoxy as well as those who mount the revolution; after all, someone must occupy the tumbrils. The year of *Pelléas et Mélisande*, 1902, also saw the premieres of Massenet's particularly Massenétique entry *Le Jongleur de Notre-Dame* and Francesco Cilea's risible costume melodrama *Adriana Lecouvreur*, both far outdated by Debussy; sixty years later, *King Priam* greets the dawn while *Opéra d'Aran* salutes the sunrise. Although German opera, along with English opera, is unquestionably at the forefront of sophistication in the writing of new works, many of the recent successes in Hamburg, Vienna, or Zurich are of a conservative cast in both the shape and voice of the libretto and the thrust of the music.

Aribert Reimann's *Melusine* (1971), for instance, not without its level of comedy in Claus Henneberg's text, revived the tale,

familiar to romantics, of the water nymph too chaste not to be
dangerous to the human male, and Wolfgang Fortner's *Die
Bluthochzeit* (*The Blood Wedding*, 1957) and *In Seinem Garten Liebt Don Perlimplin Belisa* (*Don Perlimplin Belisa Loves
in His Garden*, 1962) do for García Lorca what was often done
for playwrights in the past, yielding just enough music to get by
with. Yet even here the sway of the profane—of modernism—is
evident, in *Melusine's* coolly understated musical humor and in
the long patches of spoken dialogue that Fortner leaves gaping
out of the partitur in his stage works—so much so that every so
often a major role must be taken by an actor. Similarly, Fortner's
Elisabeth Tudor (1972) owes its bloodlines to the historical
melodramas of the nineteenth century, including the antihistori-
cal scène-à-faire for Elisabeth (soprano) and Mary Stuart
(mezzo), but again the musical style owes its biotype to neoclassical invention, and the way from Donizetti to Fortner is obscured by electronic tape, percussive derailments, jazz pastiche,
and vocal kinkiness.

Needing the casuistic discipline of atonalism but not willing
to do without the more accessible vocal language of the past, composers have taken to cutting their aesthetic with infusions of
diatonicism and impulsive anachronisms, thus further developing
the self-conscious perceptions of the modern opera score. Paul
Dessau spoke up for this approach in two reverent settings of
plays by Bertolt Brecht, *Die Verurteilung des Lukullus* (*The
Condemnation of Lukullus*, 1951) and *Herr Puntila und Sein
Knecht Matti* (*Lord Puntila and his Servant Matti*, 1966). *Puntila*, a comedy, is much the more advanced of the pair; *Lukullus*
struts right back to the jazz era, with whole scenes sounding like
pure Weill and the general mood of the text recalling the savage
Brecht of *Die Dreigroschenoper*. Actually, Dessau's *Lukullus* is

a remarkable work, totally superseding Roger Sessions' version, *The Trial of Lucullus* (1947), and transcending Brecht's shrill Marxist claptrap with music of forthrightness and power. In twelve scenes, each in a different style, the dead Roman general Lukullus is followed from his funeral on to judgment in the next world, where a Brechtian Central Committee gives him a hearing on his earthly outrages and, needless to say, purges him "ins Nichts" (into nothingness) for his crimes against the people. The first scene, Lukullus' funeral cortege, tenders a picture of brutal vividness as the catafalque and a frieze recording highlights of the general's career is dragged past the usual Brechtian mob of impervious mutterers, and the usual Brechtian narrators lecture the audience. The close of this segment, with the narrators' speeches heard over precipitous choral gibberish, possibly gives the work an artistic urgency that overshadows its thematic urgency, but a few minutes later Lukullus' soldiers bid him farewell with locker-room impiety set to what sounds like a revision of the Kanonensong from *Die Dreigroschenoper,* and *Verfremdungseffekt* is back in the saddle. *Lanzelot* (1969), another leftist philippic, proved a setback for Dessau, a fairytale satire wherein Lanzelot, a noble worker, conquers a dragon representing mass-media capitalism. Luckily, the composer regained his form for *Einstein* (1974), a sort-of biography of the scientist and as typical a work of contemporary opera as one can find. Karl Mickel's libretto veers from sympathetic historical narrative and tragic heroism to cavortings and gallimaufry, even unto the extraneous—or is it, after all?—interference of Hans Wurst, the Punch of German farce and a figure diametrically opposed to an Einstein in intelligence, philosophy, and achievement.

Facing protagonists down with their negative images is one of the theatre's longer-lived wonts, but it became a specialty of modern drama, and accordingly of modern music drama, where the counterpoint of tonalities enhances the counterpoint of words. With burlesques, bizarre irony, disorienting intermezzos, and the tempo of comic play crashing in on the drama, proper comedy was left to reclaim the *gai primitif* of the ballad comedies, as in Joseph Haas' *Die Hochzeit des Jobs* (*The Marriage of Jobs*, 1944), in which one Hieronymous Jobs fails his law orals for answering with too little pedantry in German instead of Latin and is almost married off to a nightwatchman's eager widow, to the dismay of Jobs' darling Kätchen. Modern influences may have been as common, and persistent, as house flies, but Haas' scheme is the one that has served musical comedy for centuries, with snatches of spoken dialogue, closed forms, transparent innocence, and diatonic stolidity ragged by an occasional "wrong note." Bright C Major starts and ends the show—even accompanies Jobs' examinational disaster—and not only does the nightwatchman show up in time to turn his widow back into wife and get Jobs off the hook, but the hero is also granted his law degree and a municipal post by decree of the local governor, a perfect deus-ex-machina in the person of his sheriff. With Kätchen in Jobs' arms and unblemished tonality on their lips, all ends well, and the puppet play is worlds away in the distance.

Haas' undegenerated key relationships likewise lightened successive comedies, the most successful of which is undoubtedly Boris Blacher's *Preussiches Märchen* (*Prussian Fairy Tale*, 1952), which commences with a picturesque display of turn-of-the-century Berlin in a chorus of bureaucrats singing platitudes in combined metre as they bend over their desks. More recently, composers have discovered the utility of more advanced idioms: Rolf Liebermann tested classical farce with a restrained palette of atonal coloring in *Die Schule der Frauen* (*The School for Wives,* 1955), to Heinrich Strobel's entailment on Molière's *L'Ecole des Femmes,* adopting the ensemble constructions of Mozart's day yet without the edgy ostinatos, devious as a Jesuit, that Stravinsky used to dress Hogarth in Godot's robes in *The Rake's Progress.* Gottfried von Einem, too, let the old ways sing out untarnished by revolutionary puppeteering in *Der Zerissene* (*The Man Torn Apart,* 1964), a gleaning from Johann Nestroy, the champion Viennese playwright of bizarre and parochial abilities, as untranslatable as W. S. Gilbert or Finley Peter Dunne. Boris Blacher, here as a man of words for a change, worked Nestroy up for von Einem, and von Einem worked Nestroy for style rather than contemporaneity, though the temptation to poke behind the masque might well have caught others in this age of the funny boneyard.

The age and its premises are best viewed in the career of Hans Werner Henze, whose audacious parade through genre relates to the 1950s, 1960s, and 1970s more closely than that of any of his coevals. Some are for legend, some for melodrama, some for well-made adaptation, chronicle, or special pleading, depending on what is news and to what extent tradition has made of them its slave or its guide; some are wan traversers of the old boulevards and some tread the road not taken; some find their sum early and make it work forever, others have "periods" for critics to bind up and arrange like old letters written by others to others, for critics will see patterns and patterns will be seen. Benjamin Britten, Henze's near-contemporary, is one who found his sum and held it, at an advantageous total, for life; Henze has periods. His debut in opera takes one back to the suspense of postwar Europe, when culture, clearing up the debris of wartime, wondered what the next era was going to sound like. While von Einem chose tonality for *Dantons Tod,* Henze devised a tone row for *Das Wundertheater* (*The Miracle Theatre*, 1948) and plunged in. His source for this one-act was old Cervantes, and for the score, the new serialist school, but with a light touch in the vocal line. The libretto, too, had a fresh air about its strolling players and their wonderful entertainment, which, they announce, is invisible to all but the elite. Thus they gull a status-conscious public by collecting admissions for an empty stage.

Vanity, hypocrisy, unheaval, the mystery of the green world—such topics would furnish Henze with his agenda, for thematic opera rather than idle romance attracts him. Another one-act, *Das Ende Einer Welt* (*The End of a World*, 1953), written for radio broadcast, deals with the passing of an old order of salon-occupiers, and Henze's Manon Lescaut, *Boulevard Solitude* (1952), updated Prévost to a dreamy netherworld of the present-day demimonde to emphasize destructive social forces not found in Auber's, Massenet's, or Puccini's frames of reference.

Renewing atonality with lengths of diatonic chording and knots of cadential harmonics, Henze proved what many had suspected, that circumspect twelve-tone formality is best reserved for the odd masterpiece, not for an oeuvre of dimension, and that cantabile is as necessary to the practice of opera as cyclic extravagance and the consolidation of form. Outward seeming is nothing enough, but so is a permanently obscured foundation, and a sojourn in Italy did for Henze what no study of Schoenberg could do—advise him on the continuity of canto.

Most of Henze's contemporaries have had to make the same discovery, for the European avant garde of the twenties and thirties took opera as far from the romantic era as the form ought to go, and the years since have seen a constant struggle to recon-fess the sinful traditions of the nineteenth century in modern language and structure, with much distillation of neoclassical satire within the familiar borders. There is nothing for today's avant garde to express now but the anarchic hoax, to jabber around the idea of music and text as if getting rid of them were a step in some direction or other. The composers and librettists of genuine ability have had to retrace, to compromise, to reconnect themselves to continuity.

Accordingly, Henze's next opera, his first major effort, showed the influence of sheer *singing* on the forward-seeker into atonalism. Heinz von Cramer presented him with a dazzling and complex libretto wired out of, and back into, Carlo Gozzi's *Il Re Cervò*, and with *Der König Hirsch* (*The King-Stag*, 1956), Henze entered the arena of modern myth in a sacred allegory manipulated by profane expedients, Wagnerian in length and so fully orchestrated that most of the text failed to come through. A revision in 1962 filtered out a lot of the instrumental gauze but also cut the work by a quarter of its length, deleting too much of Cramer (and Gozzi) and robbing the piece of its fleet and fidgety mystery. More about *Der König Hirsch* awaits in the final chapter; for now, let us continue this review of types, for seldom is there so convenient a revelation of epoch as afforded by Henze's output, in these adaptations—but of what sort?—and these particularly original librettos—but written for what purpose? *Boulevard Solitude* had made no attempt to "see" the late eighteenth century on its own terms, but in *Der Prinz von Homburg* (*The Prince of Homburg*, 1960), to Ingeborg Bachmann's reduction of Heinrich von Kleist's play, Henze captured much of the spirit of Kleist's hallucinating idyll, despite his insistence that productions of the opera avoid suggesting the Prussian setting. Kleist's hero—who gets his daydreams of love and glory so confused with his real life that he disobeys an order not to charge in battle, wins that battle with his charge, and then faces execution for insubordination—is a man of the era of Byron, not of Godot, and Henze scored the opera for the lush, moody bemusement of the prince, not for any modern gloss on it.

Then to chamber opera, Henze, via an English libretto by W. H. Auden and Chester Kallman, in *Elegy for Young Lovers* (1961), a stark raving vortex of characters who exist, apparently, to be encouraged, toyed with, and destroyed by a solipsistic poet concerned only with his art and how best to infuse it with real life. A self-admissive work, perhaps (Henze's *Doktor Faust?*), *Elegy for Young Lovers* redoubled the interest in the Word first sponsored by *Der König Hirsch* and *Der Prinz von Homburg*, and introduces a note of autobiographical guilt that reminds one more of the self-shriving romantic poets than of the authoritative self-dramatizations of *Palestrina* and *Capriccio*. In asking Auden and Kallman to pen his text, and on an original premise, Henze was assuring himself more of poetic commodiousness than of having to accommodate poesy, and in this he is truly modern, wanting drama of era, wanting it new and of its time, wanting the embarrassments and retaliations of today to dance on his stage. The baritone lead of *Elegy for Young Lovers*, Gregor Mittenhofer, typifies a certain form of creator—to Henze's mind, no creator at all, for his inventions are nothing less than parasitical cuttings from the garden of life, from other lives. Dwelling in a hotel in the Swiss Alps, where he is surrounded by human inspiration, Mittenhofer gets the idea for his elegy from a mad widow who has a vision of two young lovers dying in the snow. Wanting truth, he contributes to just such a death, which he can then celebrate in the elegy of the title; we see him in the last scene reading his masterpiece at the height of his elegant career.

A small-scaled work for soloists only, on stage and to an extent in the pit as well, *Elegy for Young Lovers* returned Henze to dodecaphony, though with a vocal line that discourages the jack-in-the-box intervals left over—and not meant to be—from the days of expressionism. This Henze-Auden-Kallman collaboration is a word-piece in the best sense, syllable-bright and yet amazingly musical, and will doubtless end as one of those entries that, like *Arabella* and *Jenůfa*, eventually catch on as people get the measure of it and grow comfortable in its corners.

No such acquisition of taste is required for appreciation of Henze's *Der Junge Lord* (*The Young Lord*, 1965), however, for this comedy is as endearing as anything of its age, arrayed in the costume of period but, underneath, fleshed in the skin of reprobate black humor, a true satire, and not unlike but much better in style and point than many comic operas of late. There is some resemblance to the "emperor's new clothes" joke of *Das Wundertheater* in Ingeborg Bachmann's libretto, suggested by a snippet in Wilhelm Hauff's *Der Sheik von Alexandria und Seine Sklaven,* for the young lord of the title—whose rank enthralls the gentry of a small German town so completely as to prompt its young women to adoration and its young men to . . . ah, ape his outrageous manners—is ultimately unmasked as a monkey. A surprisingly tonal score, filled with the detritus of classical buffo (including a quotation of the janissaries' chorus from Mozart's *Die Entführung aus dem Serail* when the young lord's uncle, Sir Edgar, arrives in town), *Der Junge Lord* is another of those works of today that play debtor to the shapes and attitudes of the Punch and Judy tradition, the bright outline, and creditor to the future of comic manipulation, the crazy shadow. The first act is largely sunny, all for the delight that the citizens of Hülsdorf-Gotha take in the honor bestowed upon them by the presence of Sir Edgar. For some reason he snubs their advances, preferring the company of a dingy circus troupe; at this insult to

the town, the mood changes, and Act I ends with a lumpish percussion period during which some men stealthily paint the word
"shame" on Sir Edgar's housefront. Act II takes up the friction,
for while the English visitor now welcomes the attentions of
Hülsdorf-Gotha, it is only to pass the circus chimp off as his
nephew, "Lord Barrat." This he does in music distinctly less
poised and pert than that of Act I, building up in sinister polyphony (nightmarish hoopla keeps wandering into the dainty
Tanzmuzik of the big party scene) to the exposure of the hoax
and a cathartic coda in C Major as the Kleinbürger exclaim, "An
ape! An ape!" in the darkened ballroom of the town casino, mill-
ing around, shaking their heads, a very picture of contraposition.

Having triumphed in the exactness of comedy, Henze next
turned to the extravagance of myth in another collaboration with
Auden and Kallman, *The Bassarids* (1966). This reasonably
faithful adaptation of Euripides' *The Bacchae*, one of the present
day's glories, joined *Der König Hirsch* in undergoing revision
after its first performances, apparently on the assumption that
when critics and public can't "get" a work of complex and poetic
structure the first time around, it is the work and not the restless
humans that gives offense and must be altered. But if a composer's first thoughts are not invariably his best ones, a critic's are
as useless as those of a Vandal—they *are* those of a Vandal—and,
as with Hindemith's *Cardillac* revision, it is the original text that
is to be preferred.

In fact, one of the cuts that Henze was encouraged to make in
The Bassarids was "The Judgement of Calliope," a profane rococo intermezzo plopped into the middle of Theban tragedy.
King Pentheus confronts Dionysus, the former's order against the
latter's anarchy, face to face unalterably, and suddenly a few
scenic drops descend on the outrageous laughter of the Bacchan-

tes, and Agave, Autonoë, Tiresias, and the Captain of the Guard trip on to play Venus, Proserpine, Calliope, and Adonis in a charade rampaging with pastiche and innuendo. As legend tells, both Venus and Proserpine want Adonis and submit to the arbitration of Calliope, who directs the lad to abide with each of the ladies for a third of the year, with one-third remaining him for his own devices. But even while doing their little bits as pastoral antiquarians, these principals out of *The Bacchae* stay in character; they who cavort in canzone and quartettino to mandolin and guitar were only moments before involved in a momentous struggle for power, and one of them, Agave, is going to return from her next revel with her son's head in her hands.

This is the end of modern opera, this the form that centuries of ballad comedies and mythic romances had to unite in, this "opera seria with intermezzo," *The Bassarids*. Freespoken in its currency and downright lewd in its parody, it slaps tragedy's face with its "Judgement of Calliope," arias, ensembles, and ritornellos off the rack to mock the deadly earnest of poor Pentheus, the only admirable character in the action, as he tries to keep order in a land gone crazy for lotus-eating. Can he? The sum of "The Judgement of Calliope" is that when contention reaches an impasse, a judicial compromise effects everyone's way out, but this stands at right angles to *The Bassarids* proper not only in tone but in theme as well, for when Pentheus rejects Dionysus, denying the sensuality within himself, each is playing for the uttermost power, and there will not be—as there can never be when a new god comes—a compromise, no judgment by a ludicrous Calliope in some terribly amusing get-up. When the new god is here, there can only be total suppression or total revolution.

It is the latter that Henze waits on, and why, no doubt, he was drawn to the Euripides in the first place, for one need make no sociopolitical construction on *The Bacchae* to present the over-throw of present order; it is there already in the text, along with a classic bout of theatrical tension in the figures of the realist and urban lord Pentheus and the visionary Dionysus Dendrites, the "tree youth." Their central conflict points the two lines of the score, the king decisive, declamatory, straightforward, the god chromatic and elusive—few writers fail to remark his larghetto narration of the voyage to Naxos, "I found a child asleep." Says Dionysus, and he knows whereof he sings, "I follow the god, Pentheus. *You* shall."

Chroniclers must tire of riding their hobby horses before an impenetrable public just as the public must weary of hearing about alleged masterpieces that nobody seems to like, the message being that if they're such masterpieces, why aren't they popular? Well, this chronicler is going to ride his hobby horse one more time in honor of *The Bassarids*. No one has ever explained quite how masterpieces get neglected, but they do, and "popular," as Meyerbeer will be quick to tell one, is a sometime thing. *The Bassarids* is unquestionably a work of genius—which serves Euripides right, for so is *The Bacchae*—and cannot possibly continue on the fringes of respectable obscurity for very much longer, even if it is one of those "festival" works that requires much in the way of financial outlay and rehearsal time. What did the romantics leave us? And what the classicists? Here it is, all of it, in Auden and Kallman's libretto and, ultimately, in Henze's music. Forced to tell what they do on Mount Cytheron, Pentheus' mother Agave, rapt in the new religion, tells him in a $\frac{3}{2}$ movement of this time but some other place, the harp and oboe picking at each other, the clarinet trilling, a sweet, insidious song broken up into notes unmeasured. "On a forest footpath, round a far bend," sings Agave, "flowers opened their old mouths in songs too slow for sound." But Dionysus sets his followers free, and when Agave and her sister Autonoë run onstage for the Calliope intermezzo, they are cackling over the escape. " 'On a forest footpath, round a far bend,' " echoes Autonoë in a fit of giggles. "You *are* ingenious!" "My dear," replies Agave, "I thought he'd never end." And mocking Pentheus, they prepare for the charade, footpaths and flowers forgotten . . . for the moment.

If the self-motivated poet of *Elegy for Young Lovers,* Gregor Mittenhofer, was meant to express Henze's guilt as one who battens on human disaster to feed his art, *The Bassarids* expresses unguilty, revolutionary Henze, preaching the coming takeover in the guise of the dreamy flower-child god and the spaced-out adherents of his cult. The stage directions specifically call for a variety of costumes, from medieval robes to Empire frocks, but it is noteworthy that Dionysus and his followers turn up in the jeans and open shirts affected by Europe's young commandos of the left. That Dionysus' chaos ultimately supervenes over Pentheus' order in bloodbath and destruction proves how unlovely is this flower generation, and how eager it is to substitute its own peculiar tyranny for any other tyranny going. The overthrow in Euripides is that of sensuality over grace, but overthrow in *The Bassarids* is that of youthful anarchism over the establishment. The inevitability of said overthrow is a stupid enough assumption, but it is nowhere near as stupid as Henze's assumption that he will be permitted to compose music as formalistically complex as that of *The Bassarids* or *Der Junge Lord* under the regime he is so eager to see accomplished. Then, by Ned, he will have to be "popular"—no more masterpieces, then! The man who led the chant of "Ho! Ho! Ho Chi Minh!" at the spontaneously canceled premiere of his oratorio *Das Floss der Medusa* in 1968, who cadged cantata texts from a Chilean radical (*Der Langwierige Weg in die Wohnung der Natascha Ungeheuer*) and the memoirs of a runaway slave (*El Cimarrón*), and who called the world revolution "mankind's greatest work of art" makes no secret of his politics, but this is also the man who wrestled with the art-for-art's-sakeness of symphonic form more deftly than any of his contemporaries. Does he truly think that, come the revolution, he would not be forced to work in the so unrevolutionary diatonicism upheld in Soviet countries? Or that the dissentient librettos he favors would not be forbidden him? Certainly, no Communist steerage would have permitted Henze to write *We Have Come to the River* (1976), and would probably have thrown his librettist, the playwright Edward Bond, into prison.

Only jettison the costume plot of *The Bassarids* and the piece hasn't a grain of sociopolitical stunt, but with *We Have Come to the River* Henze joins up with *Votre Faust* and *Passaggio* in the "action for music." Oddly enough, Bond's scenario is rather tame for a work of that kind, being about as revolutionary as Pennsylvania's Whiskey Rebellion of 1794; its optimistic thesis is antiwar and prohuman, and its only gesture toward the left is the closing chorus, phrased in the asseverating *faux bourdon* of the revolutionary:

> We stand by the River
> If the water is deep we will swim
> If it is too fast we will build boats
> We will stand on the other side
> We have learned to march so well that we cannot drown.

These words are sung by victims of the System, victims most particularly of the aggressive warrior-emperors and their battle empires, after numerous hardships have befallen innocent people and a General, deputy of the System, begins to comprehend the injustice of the status quo, is taken off to the madhouse, blinded, and at last drowned in the metaphorical river by his fellow madmen. But according to Bond's libretto, at length the folk—rather than the more usual solo protagonist of modern drama—will overcome.

Mob though it be, *We Have Come to the River*'s cast individualizes upwards of one hundred parts in a production that calls for the facilities of the theatre rather than the cut-and-paste of jet-set opera. On a quasibare stage of insets and multiple acting levels, instrumentalists and singers cohabit, their efforts given purposeful dimension by a subtle lighting plot and occasional entrances or exits through the auditorium. Henze himself directed the world premiere at the Royal Opera House, Covent Garden, by all accounts splendidly, but for all its episodic allegory, its carefully short-circuited dithyrambs, *We Have Come to the River* is no polemic, but an opera, and a by no means negligible entry in the lists. Despite its designation as an "action for music," so often associated nowadays with tinpot "experimentalism," and despite Henze's sympathy with the tinpot experimenters, his work is invariably art; theirs, frequently rubbish. If nothing else, Henze's artistic clarity marks him apart from his alleged comrades, such as those five composers who, with two librettists, whipped up a Holland Festival-special called *Reconstructie* (*Reconstruction*, 1969), a slimey fanfaronade of spectacular decor, fleshly exhibit, tired pastiche, and imitation archetypes (Don Juan, Erasmus, Tarzan, Al Capone, and Martin Bormann figure in the gleeful show), all wrapped up in the driveling slogans and one-size-fits-all hatred of the U.S. considered indispensable for endeavors of the European liberal. Could anything of value possibly be compounded by seven different men? Where's vision, then? Where's apprehension and penetration? Where's, even, gist? Totally of his time as Henze is, nothing puts him in better perspective than such timely gobbledegook as this *Reconstructie,* for what is this avant garde in the end if it serves only to close out any communication of form, line, or thought—of anything except its own megalomania? In art as anywhere else, renewal is essential, but any progress that denies all tradition can only betray itself—it surely won't betray many paying audiences. This, above all, Henze knows; he has spent three decades reinterpreting the bequest of the past, adjudging his roots and aligning his personal objectives in a context of transference, of moving on because one has first arrived. There aren't that many operas to be written about the unnamable, the unseeable, or the nonexistent, after all.

In fact, there are none.

4. ENGLAND

Although English opera was late in instituting any native operatic continuity to speak of, an imitation of the European model was in session by the late 1880s, each season introducing a local Wagner, Gounod, or Marchetti, and a late Celtic revival sending librettists off to *The Book of the Dun Cow* and Lady Charlotte Guest's translation of the *Mabinogion* for source material. Shakespeare was a frequent assistant in the concoction of romantic dramatics; the early 1900s were to see, to name only a few, *Much Ado About Nothing* (1901) by Charles Villiers Stanford, *The Tempest* (1920) by Nicholas Comyn Gatty, and *Romeo and Juliet* (1916) by John Edmund Barkworth: comedy, dark comedy, and tragedy, all in the decaying forms of Victorian musicality. And then, as one might put it, came Frederick Delius.

Born in England of German family and trained in Leipzig and

Paris after years spent in Florida growing oranges and learning counterpoint, Delius was French, German, Scandinavian, and even American before he was British, but this sort of adulteration was just what English opera needed, even if early interest in his works was more pronounced in Germany than in England until Thomas Beecham took up his cause. Delius' sense of kinetic vitality was never sharp, but his impressionist coloration was impressive, and he had much to share on the subject of sheer style if not theatrical line. Like the postverismo generation in Italy, Delius emphasized the orchestra, but unlike them he frequently lost control of vocalism as a theatrical medium, writing shortish, intermittently characterized operas not as good of their kind as his tone poems.

Koanga (1904), Delius' third opera, is his best, finished in 1897 before a groundplan of autonomous through-composition and moony text distracted his intelligence of operatic procedure. Based on an episode in George Cable's novel *The Grandissimes,* the libretto started by Charles Keary and commandeered by Jelka Rosen proved so foolish that the work has had to be made over in less constipated English; if this revised romance between a mulatto and an African prince sold into bondage in Louisiana still conveys too much melodramatic overkill, Delius' score is a lively and haunting confection, its black folk sound as much transcended by the foundation of German symphony as is that of the inspired *Appalachia* and *Florida.* A hint of Gregorian chant merges with spiritual in the prelude to Act III, and Delius' idea of American music owes much to the strictly Latin American dances of the Gottschalk school—but then so does the music of the American south. The plot complications in *Koanga,* involving a Simon Legreeish overseer and voodoo incantation, cannot help but look comical to modern eyes, but there is some

genuine lyrical momentum, some high-and-low of action—plus the effortless entrenchment of the many black choral passages— that give *Koanga* an outline lacking in Delius' later operas, *A Village Romeo and Juliet* (1907) and *Fennimore and Gerda* (1919). One cannot doubt the loveliness of the music, but the composer deliberately conceived of "pictures" from a piece rather than pieces entire, and the later pair simply do not project.

Still, nothing in the way of a new idea was bad for English opera in the early 1900s, for Victorian music—too polite, too brittle, to content to go on—had never equaled the flair of English romantic poetry; if there was to be no Browning in opera, there ought at least to have been a Dowson or so, someone of some assertion. Delius' work was allowed to languish too long (*Fennimore and Gerda* received its British stage premiere in 1968), but there was, meanwhile, the distinctive Gustav Holst, whose Indian *Sāvitri* (1916), composed in 1908, stands as a little landmark in this context. Attracted by India but no Rousselian folklorist, Holst laid open the storehouse of tensile, thin-lined harmonic-vocal theory for this one-act chamber work, weaving it of modal strands in one urgent pattern. There is little action in *Sāvitri*—a Hindu wife saves her husband from death by facing down the demon—but no little inward movement, a quality that likewise quickens Holst's *The Wandering Scholar* (1934), comparable in scale to *Sāvitri* but a lusty entailment on a central medieval situation: young wife, unwitting husband, bawdy cleric . . . and interloping bachelor. The versatile Holst extended himself to burlesque in *The Perfect Fool* (1923) and folk pastiche for *At the Boar's Head,* adapted from Act II, scene iv of *Henry IV, Part II,* but of his eight operas none has established itself to any degree.

In fact, virtually all of the entries of the first decades of the century are no more than index listings now, except for those by composers better known for symphonic work. Edward Elgar never even got to write his projected opera on Ben Jonson's *The Devil Is an Ass*, and Ralph Vaughan Williams could do no more than Delius and Holst to institutionalize a national music drama, for his way of combing the libretto with the loose and widely spaced teeth of through-composition gave more to orchestral than to dramatic definition. As with *Sir John in Love*, the music is rich in atmosphere but short on point, the shape aiming more at the suite than the scene. *Hugh the Drover; or, Love in the Stocks* (1924) certainly has a way with its rough-hewn, wandering hero, and the one-act *Riders to the Sea* (1937) respects the dignified despair of John Millington Synge's fisherwives and mothers, but still the current of temporal communication runs with too little discrimination. No, what was needed in these several decades between *Koanga* and *Peter Grimes* was some rightening neoterism, something to convey tradition in a cart for moderns. Vaughan Williams had his final chance in *The Pilgrim's Progress* (1951), after John Bunyan; but again the pastoral flow prostrated what might have been a curious and inviting morality pageant.

By then Benjamin Britten had made himself known in *Peter Grimes*, *The Rape of Lucretia*, and *Albert Herring*, and at last opera had made its turn in England. Not before the 1960s were other composers of rank to join Britten in profusion, but if the expedition to the peak was centuries in the ascent, the summit got rather crowded rather quickly, forced upward by the emergence of a bright generation of acting singers and guided, or at least not sabotaged, by the most eminent pool of music critics in the Western world.

Other than Britten, there was not all that much to write about in the immediately postwar years. Shortly before *The Pilgrim's Progress* came a ridiculous collaboration from Arthur Bliss and J. B. Priestly, *The Olympians* (1949), honing in on a midsummer eve when the old Greek gods, now reduced to trudging the landscape as a troupe of rinky-dink theatricals, can reaffirm for a night their former powers. A promising enough premise, this (and, incidentally, a reversal of W. S. Gilbert's premise for *Thespis*), but here it is played out in disused pre-Raphaelite masque and pantomime frivolity. As late as then, English composers were still respecting long-departed shibboleths of formula: Lennox Berkeley was taken to task for his too-episodic *Nelson* (1953), and while William Walton could not be faulted for the musico-dramatic trajectory of *Troilus and Cressida* (1954), still this arch-conservative work barely held the line at a time when revolution was called for.

Troilus and Cressida's libretto, by Christopher Hassall, borrowed some of Chaucer's details but mainly went its own way, granting an impressive opening for Walton to set for a chorus of Trojans obsessively praying to the "virgin of Troas," and granting as well the appropriate arias and duets to be expected in this subject, but appropriateness in this context is the ample tinder that somehow never catches fire, the hollow rightness. This first chorus, for example, commences in an adagio movement after a few bars of hushed prelude, quarter notes in triplets and twos preparing the scansion of "virgin of Troas" so that the five group entries come in as if out of unconscious incantation. Building the block of sound in this pattern, Walton provisions a presto section on repeated cries of "Pallas, awake!" as some bystanders wander in and bewail Troy's ruin in an angular descant. But when the doors of the temple open and Calkas comes out—"Trojans, for shame! Are ye beasts or men?"—the familiar tread of other priests and other besieged cities patters in. So it is throughout the opera: the right ideas, never quite ignited. Neither Troilus nor Cressida dominates as a vocal being, and the denoument comes off as anti-climax after the fervent love music of the second act. Of it all, only Pandarus' flighty part stays in the mind.

There is some reason now that the wheel heaves over for Britain; perhaps there is some correlation between the rise of English opera and the resurgence of English drama, although in truth no dramatist can teach a composer his art. Yet by the end of the 1960s a national opera tradition is suddenly detected, in ruddy good health and owing no more to Debussy, Stravinsky, and the Viennese school (somewhat more to Strauss) than what would be minimally polite. Britons tend to date the renaissance from 1945, when *Peter Grimes* had its debut, but Britten stands alone; a likelier jumping-off place would be the premiere of Michael Tippett's *The Midsummer Marriage* in 1955, though at the time the work did not thoroughly register with press and public—and though, as far as that goes, Tippett, too, stands alone.

The author of his own librettos, Tippett obviously agrees with Wagner, Mann, and Hofmannsthal that opera is the superior vehicle for myth, for two of his three such endeavors burrow into our preurban awarenesses for archetypal pageant and psychodrama, and the third engages gods and heroes of Greek legend. Tippett retains a solid grip on the helve of tradition, but his blade aims upwards, and now that *The Midsummer Marriage* is accepted as one of the classics of English opera, it is helpful to recall that it was thought impossibly wild-eyed and daring when young. Using the two-loving-couples—one-celestial-one-earthy syndrome that dates back to rescue opera and beyond to commedia, *The Midsummer Marriage* tries the maze of spring fertility and rebirth rituals, raising the consciousness of its two couples in a ceremonial green world for which Jesse Weston might have prepared the gloss.

This is simply said, but in truth the many doings on stage during the work's progress are so unnatural—even while acting as natural as pie—that some have found it easy to resist Tippett's collation of mythic trials. Neoclassically allusive, though romantic in perception, *The Midsummer Marriage* follows the events in a woodland haunt where people of the real world touch base with figures out of the Western superego. Jenifer and Mark and Bella and Jack are the focal couples, harried by King Fisher, Jenifer's businessman father; a He-Ancient and She-Ancient and Sosostris, an Erda who cares, represent the Beyond come Here to run tests of self-knowledge. Tippett is demonstrating how close even we moderns are to the ritual aperçus of folklore—how effortlessly the mythic passage reveals itself in the maze of machine-age life. But if King Fisher does not pan out as a Fisher King, and if Jenifer and Mark are too easy to confuse with Guinevere and Mark, with whom they have little in common, the opera works beautifully as a musico-dramatic adventure. As with *Die Frau Ohne Schatten,* one may leave it to the public to decide whether Tippett's midsummer fantasy teaches its text, but there is no doubt that it teaches as opera. Finely melismatic phrases and resourceful orchestration are boon enough; what makes the piece a permanent trophy is its canny agility, its unbroken breadth from period to sequence to development, so well measured that the action proceeds in that specificity that has no limits, the rules that bend. If *Fennimore and Gerda* was too "musical," *Sir John in Love* too general, and *Troilus and Cressida* too proper, *The Midsummer Marriage* is nothing but opera.

Having succeeded in the romantic wing, Tippett proceeded into classical environs for *King Priam* (1962), which bids fair to be one of those flawed operas that outlasts its time, its flaws forever being cited and its wisdom forever touching the soul. Only a minority opinion would claim it as Tippett's masterpiece, but its straightforward libretto certainly doesn't suffer from the mythic cullings of *The Midsummer Marriage* or the psychological gaffes of *The Knot Garden* that boggle the minds in the stalls. All is tragically open in *King Priam* and, as of old, the tragedy is brought about by a strategic defect in character; it is Priam's defect, of course—for he is the king—and it is mercy. Warned that his newborn son Paris will cause his death, he orders the babe destroyed, but the child is secretly saved, and when Priam gets another chance to protect himself, charity gets the better of him. The rest, naturally, is inevitable.

"Simply the thing I am shall make me live" is the epigraph of *The Knot Garden*. "Simply the thing I am shall make me die" might be the motto of *King Priam,* and the weighty drama moves to the inevitable on the useless advice of narrators and with the figures familiar from Homer making the futile gestures of savages warding off an earthquake. But there are heroes among them: at the end of Act II, when Patroclus goes off in Achilles' armor, Hector slays him, and then Achilles appears, Achilles himself, bellowing his monstrous, rococo war cry, echoed offstage by the male chorus, horrible to hear. Hector, Priam, and Paris stand downstage transfixed, staring out at the audience, hearing, horrible, and the din rears up too in the orchestra, louder and louder, but not ending, just shrieking to a halt all at once, and in the silence the curtain drops.

Even Tippett's detractors would have to credit his enormous finesse in orchestration. Few operas benefit from such variety in mood and texture as does *King Priam,* and Tippett is most persuasive in writing vocally for solo instruments or solo groups (as

he is in his configurations for the voice), as in the opening of
Act III, eleven measures of melody taken by the cellos. Achilles'
aria, "O rich-soiled land," accompanied by a guitar, is something
of an old master already, but so ought to be the dazzling opening,
trumpets on stage and in the pit, pounding tympani, and the
wordless offstage chorus, all giving way to a spot of light on
Paris' cradle: the thing that Priam is shall make him die.

Graduating himself into nonlinear narrative for *The Knot
Garden* (1970), Tippett created his shortest opera, one so compressed that an extracurricular cycle, *Songs for Dov,* was necessary to fill out the psychological evolution of at least one character. There are seven people in *The Knot Garden,* whirling
around in an encounter of self-dramatization and -knowledge not
unlike that in Philip Barry's play *Hotel Universe.* "Simply the
thing I am," Parolles' moment of self-truce in Act IV of *All's
Well That Ends Well,* goes on to "being fool'd, by fool'ry
thrive"; there is much fool'ry in *The Knot Garden,* the modern
fool'ry of tradition-viewing, neoclassical parody, for like *The
Waste Land, Ulysses,* and *Waiting For Godot,* Tippett's opera
roots its mysteries in what has already been discovered, in allusions to Goethe, Schubert, Lewis Carroll, Virginia Woolf, and
The Tempest, which supplies a charade for the last act. In form
and function a modern opera, *The Knot Garden* quests into identity, for national opera needs discover no more national character.
It is touchy enough just steering one's personal craft.

Meanwhile, between *The Midsummer Marriage* and *The
Knot Garden,* the annals had filled with the names of a next
generation and the titles of their issue; as with Tippett, their
procedure ripped leaves out of old books for the printing of new
writ, new turns and bows—even new sounds, through the net of
which voices of the past were raised, Elgar struggling in the

grasp of the lower strings, say, or an unprepared and terrifically Celtic yoicks out of Arnold Bax. A derivation from the past, however, is less important than one's conception of the present, and this English opera demonstrated the modernist's comprehension of form and all that form's perceptions must impart. In Nicholas Maw's *The Rising of the Moon* (1970), a Mozartean tone of libretto and a Straussian orchestra neither battled nor compromised, but shared, for Beverley Cross' text walked that fine line between comedy and dark romance in a tale set in late nineteenth-century Ireland, rebellious, jaunty, warm, and sorrowful. Behold: an original opera, plundered from no prior source, planned for musical theatre, for the achievement of itself. (Did someone mention von Hofmannsthal?) Atonal in belief, Maw is a melodist, not a mathematician; the arid retrograde-inversions of serialism would hardly do for a work set so prepossessingly in County Mayo as is this one. Martial tunes and "The Wearing of the Green" slip in and out, and at the close of the story, when the occupying English vacate the scene, a parish priest delivers a gentle salute to Ireland in Irish terms, tonal but not in the least out of place at the hem of the quite contemporary music that has preceded it. "Come you and dance with me in Ireland," he sings.

The dance was capable of winging into darker precincts, such as those where Godot calls the steps, where man dances right up to the mirror for a look and dignity is at a discount. Harrison Birtwhistle's *Punch and Judy* (1968) gave the commencement address for operatic satire in England, feeding on parody and quotation to pierce through to the obverse of reality. Here, again, is the night of the puppets—how far along they have come from mannerism and Maeterlinck, from neoclassical yet late-romantic *Doktor Faust*. In Stephen Pruslin's libretto and Birtwhistle's end-stopped numbers, the old dolls come back to death for fun

and games, attaching such sport as Punch's murder of the interlocutor, here called the Choregos, by taking him for a cello and
bowing him lifeless. Murder runs rampant in *Punch and Judy*,
as it did in the old Punch and Judy shows, but the violence
amounts to more than silly social commentary or *Zeitoper* tactics:
it is endemic to the thrust of the masque, to the extreme of low
comedy, and even to the hair-raising cacophony of Birtwhistle's
orchestration. This is the ultimate of the profane, using burlesque, alienation, parody, and pastiche to intellectualize the ego
of modern man in a modern art. Comedy is its key, and the
wonder of it is that a romance such as *The Rising of the Moon*
shows no less vitality than a comedy such as *Punch and Judy*—
that today's ego need be neither wholly brutal nor wholly sentimental, that so much artistic validity lies between. Not till now,
perhaps, can we appreciate what a richness of temperament *Der
Rosenkavalier* radiated when new, for it has lost nothing in these
years of wild advance.

On the other hand, many current tonal constructions do seem
a bit . . . simple? There is Malcolm Williamson's *Our Man
in Havana* (1963), for instance, from Graham Greene's novel
about a timid nobody who invents a spy network only to have
the fantasy turn real on him. Williamson's idiom is aggressively
conservative, and some of his more discrete moments sound like
extracts from a sightly daring musical comedy. Yet *Our Man in
Havana* is a canny exercise in melodrama, and does surely earn
more points as an excursus on the evil manipulations of jet-age
society than does Anthony Gilbert's one-act *The Scene Machine*
(1971), a trite revelation of commercialism in pop music.

Clearly, comedy, in mode and tempo, is the particular of modern opera one way or another, and it needn't be all that grotesque
or vicious in defining its terms and taking shape. Comedy may be

contained by a larger form, as tragedy cannot—at least, it pretends it may—and even the nicety of a sharply characterized play may be commandeered by a resourceful librettist, held to its original negative in reduction, and then transformed by the musician with the alternate niceties of music. Comedy has always tended to accuracy, but it is far more accurate in a piece such as Richard Rodney Bennett's *A Penny for a Song* (1967), carefully taken over from John Whiting's play, than it was in *Don Pasquale* or *Der Barbier von Bagdad*. It is at its most accurate today, for that is what all music has been stretching for since the start of the present century. In the age of the sacred, it was expansion; now it is accuracy. For French opera, the capital of the transition might be *L'Heure Espagnole*, for Germany, *Wozzeck*—not the scene but the measure, the one moment in its fragments, and then the other movements one by one—and for England it's *Façade*, with its "new landscapes" of a "brief but memorable vividness." One recalls how sudden *Façade* felt in its time, and recalls as well the convulsive premiere of George Antheil's *Ballet Mécanique* in Paris in 1925, when the wind machine and pianolas inspired fist-fights and contumely in the aisles—but there was Eric Satie crying, "What precision! What precision!" In *A Penny for a Song*, precision is served by farce, laid on the coast of the channel during the Napoleonic Wars, when eccentricity and mistaken identity keep everyone busy for hours. Bennett's music has that recommended exactness truly enough, but the core of the opera, as with the play, is its remission of nostalgia, its refusal to let jokes and decor serve as all, its bite. This is its accuracy, and it may be that the words handle it more succinctly than the music, as when an old rakehell, now in retirement, recalls the poetry-writing of his youth, rapturous then but foolish now. "The old songs that make me cry can only make you laugh," he says. "That's what time can do—cover my face."

The English forte is words, and words are the key to modern opera. The music one takes for granted; one expects the music, and one didn't complain when the nineteenth century put song ahead of story. But now that the Word is with us again, English opera comes into its heyday, and there may be said to be a literature of librettos, a poetry certainly. So respectful of the text was Humphrey Searle in his *Hamlet* (1968) that the music failed not only to project it but failed even to project itself. "To music or not to music," wrote Karl Heinz Wocker of Searle's *Hamlet* . . . and his conclusion? "Not to music." Yet Searle had done well in other operas, such as his *Diary of a Madman* (1958), for which possibly the fact of adaptation from Nicolai Gogol's Russian left him free to express himself; in any case, his tone row is more congenial to Gogol's disoriented little clerk than to Shakespeare's too-well-oriented prince.

Joseph Conrad's *Victory*, too, became an opera by that name in 1970, in Richard Rodney Bennett's music and Beverley Cross' text, but one is on trickier ground encompassing a novel in two or three hours than one is encompassing a two- or three-hour play. What to adapt and how was never a problem in the 1880s; now that form is all, it is a problem; it is a crisis; it is a sensation. "Form is the Cage and Sense the Bird," trilled Henry Austin Dobson—nay, form *is* sense. Form is the voice through which sense makes itself intelligible—form can speak—and while form in opera was little entertained for so long in England, England is now all for opera and form is its stop in the mind. Not surprisingly, more was said for and against *Victory* on the subject of shape, communication, and source than was said on the presumably significant question of To Music or Not To Music.

Indeed, now that stage technology is so free with gadgets and jesuitical lighting design, it seems to be every operagoer's responsibility to separate theatrical display from the material itself, and now that music has become so complex to first-time hearers, it becomes likewise necessary to decide if one is only getting the words, or getting nothing but words. Bennett has his reasons for choosing a "difficult" idiom and occasional spoken lines—as if to deflate the crescendo of the action, and keep it psychologically in the interior—for the music of this literary opera *Victory*, but one notes that *Victory*'s structure falls in with by no means revolutionary thinking that obtained fifty years ago, and that in pursuing the three-dimensional version of Conrad's heroine, Lena, both Bennett and Cross did away with the very ambiguous design spun out for her in the novel . . . one might go on. The point is that the outward shape of *Victory* is not particularly new, but neither is it old, as is the shape of *Our Man in Havana,* to pick a musically old-fashioned example. Williamson's aural ethos is way to the rear guard, Bennett's somewhere to the fore (excepting the modish crazies)—and in that one takes his sample, for once a composer sets to his task, framework is nothing and the *sound* of the Word is all. The prevalent arioso of Bennett is not merely a different language from the more "number"-oriented Williamson, it is a communication of point-of-view, and one that strains close to its two focal characters and their reticent dialogue; thus their lines count all the more. Williamson's more mercurially melodic *Havana* is less rapt, less ambiguous; when his vacuum cleaner salesman's invented agents and counter-agents start to materialize, Williamson pursues the comic irony of conjunction, while Bennett's is the irony of tragic conclusion. Musical style is the decisive factor in the effectuation of operatic character, and though the Word is in its heyday now, it is yet subject to a certain bargain with the music, for the Word is dependent on the music, while music is an absolute unto itself.

And this brings us to Benjamin Britten, for no one has implanted the cut of form in English opera so securely as he, even as the contributions of his contemporaries point in vastly alternative directions. Strange to tell, his first opera took the form of an American *Singspiel, Paul Bunyan* (1941), with a libretto by W. H. Auden and a treasure trove of "native" pastiche from ballads to blues, made notorious by default because the piece flopped in its American debut and was not staged in England before 1976. It is even stranger, then, that Britten's next sally in musical theatre should have been the uncompromisingly East Anglian *Peter Grimes* (1945), not for the sudden wrench in regional perspective, but because that perspective was so brilliantly consummated. In the opera's prologue and all six scenes the overall atmosphere hovers close up to the ear, whether in orchestral coloration (the famous "sea" interludes) or in the extended lyrical episodes (the famous "embroidery" aria) or in the large-scaled ensembles (the famous "O tide that waits for no man spare our coasts!") or in the intimacy of monologue (Grimes' famous "What harbour shelters peace?"). Every scene in *Peter Grimes*, every turn of action and technique, it seems, is famous as a case in point of opera composition; what other modern work is so insular in feeling, so hostile to outsiders, and yet so necessary to know?

A chorus opera, *Peter Grimes* derives from George Crabbe's poem *The Borough,* and belongs as much to the men and women of the town as to its titular hero, the despised and despising Grimes. An apothecary, a lawyer, a schoolmistress, a tavern-keeper and her house doxies, and resident fisherfolk fill the stage with the curiosity and intolerance of the societal microcosm, all in re-creation of the predestinate naturalism of the verismo era. How changed, the mettle of romantic music, however! How altered it is even as it does what it did! The question posed, once

again, is *who?*, and the answer is *Grimes against the lot*. A brutalized, ostensibly insensitive fisherman who goes his own way
in the face of the communal suspicion of "The Borough," Grimes
is first seen at an inquest into the death of his apprentice. He explains it as a fishing mishap; the Borough calls it murder, and
while no indictment is brought in, Grimes is made the resident
pariah more than ever. So it continues for Britten's protagonist,
a very paradigm of the modern antihero, neither sympathetic
nor admirable but, for good or ill, drawn in full: mishaps, the
bad luck of environmental fate, dog him until there is no room
in which to move forward and none for retreat. On the advice
of a sort of friend, Grimes takes his craft out to sea and sinks it,
and the hateful folk of the Borough close this inquiry into mob
action looking out on the water at the report of a boat going
down in the distance. "One of those rumours," says the tavern-
keeper.

In an era when composers are strong in the technology of
opera but weak in broad melodic appeal, it is enlightening to
browse through the *Peter Grimes* score and note how much of its
appeal adheres to conceptual plan. No one would deny the effectiveness, as coup, of the full chorus' "Home! Do you call *that*
home?" at the close of the tavern scene, when Grimes bundles
his new apprentice off with him into the spit of a storm. But it
is all the more effective to hear "that home" as an inversion of
Grimes' own rising ninth (as in the first two notes of "What harbour shelters peace?"), so associated with his ambitions of success
and respectability. More accessible to the first-time ear is that
remarkable passage, "Now the great bear and Pleiades," in
which Grimes sings repeated E's over a four-part canon in the
strings, suddenly immobilized on a chord as Grimes takes the
fifth entry of the canon, "breathing solemnity in the deep night."

Unlike Hans Werner Henze, the other giant of this period in

opera, Britten never hymns anything but the self. Grimes, Lucretia, Tarquinius, Albert Herring, the Governess and her ghosts, Owen Wingrave, and Gustav Aschenbach take their own parts, the symbolical Billy Budd and Claggart of Melville's novel are much less so in Britten's opera, and even glorious Elizabeth I is herself, never glorious England, though temptation, especially in a "coronation opera," might have led any number of composers into national allegory. Henze frequently enough concerns himself with universals, lessons for everyone, but Britten is less the teacher, more the dramatist; what the two do have in common is the melodic idea, the one-almost-has-to-say romantic vocal line, and if Britten's idiom is conservative and Henze's largely atonal, the Englishman's astounding assets of invention make his music no less "unheard before" than Henze's. It is well to note that in their linear narration and fantastic or quasifantastic dramaturgy and in their motivic technique and vocal shape, Britten's operas are the culmination of the romantic era, save for a guidance by the fixed forms of neoclassicism. The developmental contouring of Britten's orchestral interludes, reclaiming the Wagnerian leitmotif from its devastation by overuse in twelve-tone opera; the palpable architecture of *The Rape of Lucretia* and *The Turn of the Screw;* the threnody in *Albert Herring,* the intersectional construction of which is all the more apparent because its very seriousness is so "wrong" for the comic climax about to descend on it; the stylistic parodies in *Gloriana, A Midsummer Night's Dream,* and *Death in Venice*—this is the substance of neoclassical order put to the shadow of romantic escape, usually into dreamworlds of private, but sometimes public, fixation. The satiric tragedies of the age of Godot link mannerism to neoclassicism; Britten's tragedies link classicism to mainstream romance.

After the full-scaled *Grimes,* Britten tried his hand at chamber opera. His first, *The Rape of Lucretia* (1946), from André Obey's play, prefigured the narrative frame of *King Priam* in miniature with a two-person chorus relating a Christian subtext to Obey's tale of a saintly Roman wife violated by a dissolute soldier. Outwardly cool, inwardly sensuous, *The Rape of Lucretia* was a study in the decocting of operatic expression, a belt-tightening tour de force of variety within the small scope. The second chamber piece, *Albert Herring* (1947) came off almost as an anomaly in its day, a jolly package of comic premise executed without a hint of the puppet masque or the tragicomic waste land of waiting. Eric Crozier's libretto, loosely based on Guy de Maupassant's story "Le Rosier de Madame Husson," follows the emancipation of a village lad who, for his repressed little life, is crowned the local May king, the local girls being found too wanton for the taste of the May committee of small-town dignitaries. Lemonade spiked with rum effects a revolution in timid Albert, who vanishes, off on a bender but presumed dead and already in the process of being roundly mourned when he turns up at the finish, a man new made.

Burrowing into the warrens of rhyme and the vernacular, Crozier devised a comic inversion of Montagu Slater's text for *Peter Grimes,* using the same closed society, local types, and misfit hero, but, of course, for the vacation of a genial spoof rather than the destruction of tragedy. Slater's fisherfolk in the Borough phrase their verities in poetry:

> Now is gossip put on trial;
> Now the rumours either fail
> Or are skated on the wind
> Sweeping furious through the land.

Crozier's tradespeople and satraps of the inland hamlet speak the crisper argot of comedy: "Hi! Hi! Heard the news? Heard the news?" calls a boy during the morning-after dragnet for the disappeared May king. "There's a big white something in Missus Williams' well!" Shortly after, the town philanthropist and beldame, Lady Billows, puts in her oar: "Modern methods! That's what we need! Bloodhounds! Fingerprints! Electro-magnets! Water diviners!"

One wants to compare Henze's *Der Junge Lord* to *Albert Herring*, but, really, there is no analogy other than the basis of farcical structure. Henze's black comedy yields black music; Britten's score for *Albert Herring* is as sunny as the April daylight on a curate's cottage. Calling a halt to the all-encompassing cyclicism of Wagner and Schoenberg, Britten turns the leitmotif from a calling card into a variation form. For example, the central theme of the opera is associated, *very* generally, with the May day celebrations; as it comes and goes, tossed from character to situation and back, it delivers unity but not specificity—that much is left to the words. Turning up first in Lady Billows' reactionary outburst, "May Queen! May Queen!," these four notes (C-G-E-C of the C Major triad in second inversion) become the fugue subject of an ensemble crowing over the crowning of a May king, turn up as the horn theme of the second-act prelude and again as bits in the mosaic of the May picnic speeches. Then, when the vicar launches his toast to "Albert the Good," the theme predominates in an altered inversion which in turn is developed in the following interlude, frames the next scene in a flute solo, and shows up at last in its original form, momentously, to lead in the threnody of mourning—sung, in the absence of a corpus delicti, to Albert's crushed may wreath, "found on the road to Campsey Ash . . . crushed by a cart."

After an intermission spent in quasichildren's opera, *The Little Sweep* (1949), Britten again resorted to the large stage in *Billy Budd* (1951), as rich and symphonic a work as *The Rape of Lucretia* was compressed and austere. Crozier and E. M. Forster made the adaptation of Herman Melville's tale of the pure fool of a seaman too beautiful for evil to suffer and, unlike Albert the May king, incorruptible and therefore not to be educated but to be destroyed. No, Billy Budd is more like Grimes, the misfit trapped by society in a state of siege, only Grimes was a bad lot and Billy is beautiful—"O beauty, O handsomeness, goodness," as Master-at-Arms Claggart apprehends him. Claggart, as evil as Billy is good, is the prize of the piece, a black bass used in contrast to a succession of tenors and the lyric baritone assigned to the hero.

More compact than *Peter Grimes* in its action, *Billy Budd* is conversely more expansive, suggesting if only to some degree the immensity of Melville's contest between purity and deviousness. The younger work has not exactly been done to death on the world's stages, possibly because of its ultra-British flavor, but it may yet join *Grimes* in popularity in the English-singing world, for its vocal-orchestral integrity makes for as richly profiled an evening as anything of recent vintage.

Interestingly, the neoclassical tenets of the Italian composer Giorgio Ghedini held *his* operatic draw on the same source, at nearly the same point in time—and also named *Billy Budd* (1949)—to one-act length. But Britten's version is a giant work, fully fledged in the apparatus of romantic surge (though lacking even a single female voice), yet so dextrous in its thematic manipulation as to cast the second-raters of the nineteenth century in a suspect light. This, after all, is what they were working for, a limitless confluence of leitmotifs serving character and idea. Why, then, did they not achieve it with such symmetry as here? Because, perhaps, their forms would be no forms until the neoclassical interlude restored the balance of shape to action, and of word to sound. Logic, at last, revives romanticism, and replies to its unanswered question.

Logic especially directed Britten's coronation pageant, *Gloriana* (1953), his most underrated opera, the processions and episodes of which take some getting used to, but do truly repay acquiring the taste for them. Covering the last years of the reign of Elizabeth I (and performed to inaugurate that of Elizabeth II), William Plomer's libretto explores the conflict between the Queen and the Woman. No heavy line divides the one from the other, for the Woman, ultimately, is the Queen, and everything in her converges to duty and greatness. Her tenor is the Earl of Essex, Robert Devereaux, as in opera, play, and movie before, and with the traditional end for the ambitious young lover. An anticipation of his fall makes for a startling second-act curtain when, after deliberately insulting Essex's wife by traipsing around in her dress, the Queen makes "Robin" her Lord Deputy in rebellious Ireland and the crowd strikes up a martial anthem, "Victor of Cadiz, overcome Tyrone!" while Essex's enemies at court rejoice that this will feed his pride and push him closer to the abyss. Sure enough, the orchestra on stage donates a coranto in D Major to the occasion, the Queen takes Robin's hand, and the pit orchestra sullenly moves in with the melody of "Victor of Cadiz" in a minor, building to a crescendo in unharmonized octaves. The puny coranto fades behind this superior commentary, and the curtain slowly falls, closing off the dance and ceding to the cavalcade of vaulting hunger.

Looking back at the other composers who stocked the repertory in fruitful recidivism, and at those as well who, with or without success, at least produced operatic entries on a regular basis, one finds that Britten alone of them all exercised his gifts at maximal level throughout his career. From *Peter Grimes* to *Death in*

Venice, there is no stunning rehabilitation of style as in Verdi or Wagner, and no all-or-nothing treatise in the manner of Busoni's *Doktor Faust*—and yet no lack of variety, no hiding behind the last. In the three "church parables"—*Curlew River* (1964), *The Burning Fiery Furnace* (1966), and *The Prodigal Son* (1968)—and in the miracle play *Noye's Fludde* (1958), Britten surrendered much of his romantic armament without fumbling for presence, and in setting Myfanwy Piper's text after Henry James, *The Turn of the Screw* (1954), he aggrandized the classical corollary by fitting the sixteen scenes to an introduction, twelve-note theme in fourths and fifths, and fifteen variations, controlling the most uncontrollable of romantic notions, the obsessive delusion (in this case, that of the Governess for the absent guardian of the children in her care). True, these chamber works are hardly an innovation for the composer of *The Rape of Lucretia* and *Albert Herring,* but nothing in the austere *Lucretia* or the bumptious *Herring* saw study for the now-repressed, now-hysterical *Turn of the Screw.* One congratulated Milhaud for his variety of mode, Strauss for his quest of the Word, Malipiero for his curious worldview, Křenek for his undying fascination with the administration of expression. But no one of them passed his journey so consistently as Britten; none was able to engage the public and stun the scholars with such security.

It is a lucky strike that Britten decided to tackle *A Midsummer Night's Dream* (1960), his rebuttal in the controversy over whether plays with so much music of their own as Shakespeare's ought to be set for song at all. The libretto that Britten arranged with Peter Pears follows the play straight along in text already known, letting only the music be the new. And this is what music can do: not hinder Shakespeare's verse or mope along under it, too respectful for words, but formalize the intermeshing of fairy, mortal, and "mechanical" worlds in what amounts to three different operas taking turns. Mendelssohn wouldn't know this forest, not even when Act II opens with—yes!—four chords, yet scored not for woodwinds in E Major but for strings, brass, wood-winds, and percussion, a chord for each and each chord remote in key from the others. This is the forest primeval, the forest of real imaginings and transformation, a modern forest. The fairies who inhabit Britten's opera are a rather contrary group, heralded by throbbing gasps from the lower strings and given to distinctly off-the-wall vocal sequences, now scalar, then chordal, while the two loving couples share a declamatory arioso, *espressivo,* and the mechanicals cavort in rhythmic recitative.

Perhaps this was the chance to tie the performance of *Pyramus and Thisby* in the last scene to the preceding action, to demonstrate in music what Shakespeare could only prove by words—that love is metamorphosis or it is nothing. And then again, perhaps this was not the chance. Britten's carefully constructed little bel canto parody bears no relation to the "translations" and enchantments of the forest, where not only the mortals but even the fairy queen learn that love is best contrived by transmutation, that seeing one's beloved through a chink in a wall, as Pyramus and Thisby do, is seeing too clearly: it is not love. Still, the authors left Shakespeare's lines in to teach the ruse, and Britten underlines the most telling line of all in a fugato quartet for the two couples in Act III—"mine own, and not mine own."

Like many of his colleagues, Britten relied on adaptations, but—and this is of interest to moderns—his sources inclined to prose rather than drama. *A Midsummer Night's Dream* is his only borrowing from other stages for a full-length work, and thus Britten encountered none of the problems that have beset Penderecki, von Einem, and numerous American composers, that of finding the musico-dramatic transmission for an organism already profiled for drama. Britten has been lucky in his collaborators; his colleagues too frequently accept the stony assistance of the Word written for speech, which is no collaboration, not for opera. That Britten and his librettists were masters of whatever it was they were up to each time they did it, was inadvertently brought out by Peter Maxwell Davies' *Taverner* (1972), introduced between the premieres of Britten's last two operas, *Owen Wingrave* (1971) and *Death in Venice* (1973), both scripted by Myfanwy Piper. Davies wrote his own libretto for *Taverner*, based on the life of the English composer and ecclesiastical politico, John Taverner, and while it would be improper to compare Britten and Piper's formats to that of Davies, it is inescapable that Davies' music simply doesn't transmit ideas as well as Britten's. *Taverner* works as theatre, certainly—as an experience, but not as an organism; there is no growth in it, no inner definition, just the pictures and patter in the postexpressionist style that Davies had utilized more effectively in the smaller terms of his *Eight Songs for a Mad King*.

How much more of an interchange between stage and stalls was there in *Owen Wingrave,* after Henry James' story about the scion of a military family opposed to war but not to acts of courage, such as tangling with a family ghost. Not only was Piper free to create her eight principals (no chorus) from the "flat" dimension of James' text, but she was also working for a queer commission, in the round—a debut on television and a form for the camera. More elastic than the stage at shredding and restitching the hems of time, television seduced *Owen Wingrave's* authors with a few telefunctions, but still the case was made by their expertise in plain old musical theatre. Here, again, was an opera, whether on screen or stage, its music sculpting the action instead of—as with *Taverner*—getting in its way and occasionally even corrupting it.

And here was another opera when Britten and Piper turned to Thomas Mann's *Death in Venice,* as thoughtfully layered a source as James' story and much more visually symbolic but captured in full in music and text. Mann's hero, the writer Gustav Aschenbach, ended up as one of the most developed characters in all opera, set off by the seven baritone incarnations of his evil genius, who materializes in the first scene as the Traveler—"from beyond the Alps by his looks"—who tells Aschenbach what he must do and then reappears in the second scene, on the boat to Venice, as the painted Elderly Fop, showing Aschenbach what he must become in the languid and scrofulous city of the lagoon. Later, as Dionysus, his voice fights with a counter-tenor Apollo for right of way in Aschenbach's *bildung*—a central conflict in Mann's work, the beast of art in contest with the order of life. Thus the original story comes complete to the stage, and ideas do not die.

Aware of time passage, Britten turned modern in his farewell to art: the leading themes in *Death in Venice* are molded by half-step intervals, more for the row than for the tonic triad. One had hoped for a faultless farewell—such is the arrogance of a spectator—but perhaps the finale of Act I, a pentathlon on the beach as "the feasts of the sun," does not come off, given over as it is to a squad of little boys *en cabriole.* Moreover, it just may be that Aschenbach's metaphorical obsession with his stripling becomes too literal on stage, ungainly to see as it isn't to read. Never mind; never mind. There was a wonderful expectation in viewing Britten's last opera (for one knew it was the last) that is seldom granted to today's opera buff, now that the continuity of the repertory has sagged to the formality of dutiful premieres—

dutiful critics, dutiful public, dutiful works. Events we do not lack, but the reason for the event is sorely missed, and Britten's output was the last such reason to date—not a trace of the dutiful about that adventure. Expectations had every right to be high for *Death in Venice* and, barring one's personal quibbles, expectations were met to the nth. "Where," to reword Katisha, "shall we find another?"

5. THE UNITED STATES

Americans were the last to realize a national opera tradition, the last to release romanticism from its contract, and the most squeamish about facing down modern art—odd, for a people so frequently accused of embracing novelty and junking acquaintance at every opportunity. The explanation for America's tardy enthusiasm for American opera is not easily come by, for the production of Donizetti, Meyerbeer, Gounod, and Verdi was an ordinary feature of city life east of the Mississippi in the nineteenth century, as Colonel Mapleson's memoirs attest, and the fame of the Metropolitan Opera in New York tends to obscure for us now the almost reckless activity of the less distinctive companies that bestrewed the map in the early twentieth century, frequently with homegrown and locally cast pioneer efforts in the taming of music drama's particular manifest destiny.

But the sad truth of the matter is that, for all the eagerness to write it, attend it, or be in it, American opera is still an elusive art; few are the American operatic masterpieces—few, even, the mere first-raters. One reason for this, often quoted but no less resonant in the repetition, is that America does have a much more vital popular musical theatre than any other nation—by way, doubtless, of putting the principle of democracy into artistic operation, *e pluribus unum* or so (the *opéra comique,* antecedent of the musical, was likewise the favored form in revolutionary France)—and the prolificacy of the American musical has channeled energies that might otherwise have concentrated on "the real thing." Trained musicians, from George Whitefield Chadwick and Victor Herbert to Marc Blitzstein and Leonard Bernstein, have asserted that the commercial theatre is the place for America's national opera, and they did not have ramshackle musical comedies in mind, though in all fairness it must be admitted that Chadwick's *Singspiel, Tabasco,* proved one of the more ramshackle begettings of the year 1894. Opera has been, if not a staple, a regular visitor to Broadway, long before *Porgy and Bess* and long after; Britten's *The Rape of Lucretia,* for instance, turned up at the Ziegfeld Theatre in 1948 for twenty-three performances—about fifteen more than the average opera would get in any single season. And one must concede that the salient characteristic of the musical, from Adam de la Hâle's *Li Gieus de Robin et de Marion* in the thirteenth century right on through *Die Zauberflöte* and up to *Pacific Overtures* in 1976—its insistence on dramatic procedure, on the directness of the profane and the immediacy of the vernacular—is a quality that opera, in its more fulfilling musicality, has often admitted.

There is another reason for America's too-slow maturity as an opera center, this one less frequently advanced, if advanced at all; it is time to advance it. This is a matter of the influence, a most pernicious one, on an impressionable public at a time when it is ready to receive opera but is in need of some education. Said influence has been wielded in the post-World War II years by a platoon of émigrés, most of them from Vienna, who have as translators, lecturers, and small-time impresarios treated the operagoing American as an imbecile who must be led about in a kind of kindergarten from Mozart to Puccini as if from milk and cookies to the pledge of allegiance. This has been the era in which opera opened up on these shores as a modern community concern, a place of pride in regional culture—yet this was the era of puerile background seminars and English translations that turn Da Ponte and Illica into loathsome nonsense. It makes those without sufficient knowledge see opera as great music wedded to doggerel; they are likely to turn their attentions to more reasonable matters.

With that sort of education gulling the audience, no wonder contemporary American opera lacks such luster: what can the creators do for a public so misled? But as far as that goes, what can a public do with creators who mine the nation's vagrant operatic tradition when said tradition consists largely of shopworn romance and sterile adaptations? Tradition, is it? Those American operas of the early twentieth century that have survived for scrutiny do not survive the scrutiny; they warble, squeal, and pine away in a variety of late-romantic imitations, and the occasional native voice that makes itself heard above the din is no mitigation. For the romantics there were Frederick Shepherd Converse and Horatio Parker, just for example, whose *The Pipe of Desire* (1906) and *Mona* (1912), respectively, showed in both cases the results of too much study in Munich. *The Pipe of Desire* (Scene: a woodland; Time: spring) was no worse than most of the *Märchenopern* then popular in Germany, and *Mona,* a dip into Saxon Britain with a chorus of "Soldiers, Druids, Bards, and Britons" and a theme antithetical to that of *Gloriana* (Princess Mona learns that by denying her womanhood she has ruined rather than ruled her country), actually scored something of a success at its Metropolitan premiere. But even so, these were hollow entries, cut raggedly of other men's cloth, while the native voice, such as that of Charles Wakefield Cadman, fared no better. Cadman was into the folklore of the red man, and had logged time in procuring Indian modes on the reservation; this duly resulted in *Shanewis* (1918), subtitled *The Robin Woman*, and *The Red Rivals; or, Daoma*, which for some probably very good reason was never professionally produced.

Some hopes rode on the appearance of Victor Herbert's *Natoma* (1911), if only because Herbert was the master melodist of operetta at the time, but again a leaden process killed the possibility of anything more than a passing success. Here, as elsewhere, Mary Garden gave her all for the cause as Natoma, the homeless Indian maid who murders a villain during a frenzied "Dagger Dance" and endures a certain justice by entering a convent, and because Joseph D. Redding's libretto gathered together Americans, Indians, and Spaniards in old California it was thought to be a veritable token of American portrayal. Unfortunately, while Herbert had no trouble negotiating the musical end of things, if only to competence, Redding's text sabotaged the endeavor, particularly his Gitche-Gumee lyrics for Natoma's big narration in Act I, "Would you ask me of my people" (which goes on, "of my father and his father? Then I bid you now to listen"). Herbert set this scene, unlike Bartók and the Gitche-Gumee text of *Duke Bluebeard's Castle*, without attempting to relieve the sameness of the scan. Worse yet, he closed Natoma's grand outburst with a repeated f sharp minor seventh chord, that noise which has since lived a false folklore on the soundtracks of western movies.

The 1920s should have brought some insurgence in American opera as it did in American music, whether popular, serious, or that netherworld, invested by both, that stands so paramount in

American art. But besides the jazz-tainted works previously listed, Louis Gruenberg's *The Emperor Jones* and George Gershwin's banal one-act, *Blue Monday,* the interwar years at first had only George Antheil's *Transatlantic* (1930) to spark something in the way of an American opera—and that in a work suspiciously similar to *Jonny Spielt Auf* in form and content. An almost nonstop barrage of tango yowls, syncopation, and flatted thirds, with the libretto (Antheil's own) set so that the vocal line accented the wrong syllables of words, *Transatlantic* followed the progress of an American presidential election, making passes at a statement on political corruption and ending up, like *Jonny Spielt Auf,* as a celebration of the breakdown of the romantic era and its sacred intentions, spoofed by the jingly piano and the mooching saxophone. An anarchic vaudeville hoofer and a bathtub vignette upheld nothing but mockery, and old American campaign songs were quoted for further alienation in a scene in election headquarters, while the chorus chanted nonsense about big business, the machine, and money, and three policemen rushed in to query the candidate's views on the Volstead Act, to the tune of "There'll be a Hot Time in the Old Town Tonight":

> Do you stand for better laws for the right?
> For Prohibition's our delight!

Jonny Spielt Auf was heard at the Met; *Transatlantic* was not—perhaps because one such exploit was considered sufficient all around—and not only did the piece come to light in Frankfort-am-Main rather than at home, but the score was published by Universal, a Viennese firm.

Dumb as it was, *Transatlantic* was the exceptional work, and the rule was still tacking, when it tacked at all, to romantic formula. The momentous Walter Damrosch—no Indian one-acts for him—was silent throughout the dashing twenties, having already donated his *Cyrano de Bergerac* in 1913 and not being heard from again until *The Man Without a Country* in 1937. Where was the true native spirit, however regional, to define the land in sound and plastique the way *Sir John in Love* did England or *The Excursions of Mister Broucek* did Czechoslovakia? Where were our entries in the sweepstakes of experiment in Western music drama? (Alas, Charles Ives never did write his opera-to-be, *Major John André.*) It is true that much American art is blocked by its own sectionalism from trading in the marketplace of European tradition—fine! all the better! Let modern opera abroad fly barriers in common commerce; give us the wisdom of locale, whether urban or rural.

Wisdom at length we got, and the first two American operas of merit, *Porgy and Bess* and *Four Saints in Three Acts,* relate as national treasures in the narrowest sense: they are too individually derived to suit a continent, *Porgy* in its Charleston ghetto and partly popular style, *Four Saints* for its Baptist dignity and modish libretto, a southern fish-fry held at 27 rue de Fleurus. They are originals, is what they are, all the more so for having entered the scene on the heels of Deems Taylor's *The King's Henchman* (1927) and *Peter Ibbetson* (1931), the latter a dreamy nineteenth-century excess and the former, versed in archaic but intelligible Anglo-Saxon poesy by Edna St. Vincent Millay, given to uncomfortable resemblances to *Tristan und Isolde* when it wasn't resembling Dvořák, Mussorksky, and Humperdinck.

No, *Porgy and Bess* and *Four Saints* turned the game inside out: they were themselves. Produced a year apart in the mid-thirties on Broadway with all-black casts, both survive as vital experiences in American music. Virgil Thomson's *Four Saints* ap-

peared first, in 1934, "an opera to be sung," albeit in Gertrude
Stein's idiosyncratic libretto (has there ever been a more ear-
catching opening in opera than Stein's first line, "To know to
know to love her so," set to Thomson's tonic-dominant oom-pah
in F Major?) about absolutely nothing at all. Actually, there are
four acts, or so the characters insist, and rather more than four
saints, and the diabolical naiveté of the proceedings is fulfilled
in Thomson's hymnlike episodes, choral antiphony, plagal cadences, and diatonic glee. *Porgy and Bess,* of course, is a much
more complex work, rich in atmosphere, character, and plot and
hovering, like so many of its successors, in the air between the
Real Thing—or what one is used to calling Real—and the presumably less dignified musical theatre. *Porgy* and *Four Saints*
have this much in common, however: they were operas, they
were American, and they founded a school of *sui generis* biases
that, for a moment, expressed something percipiently American
instead of hacking out dead Continental melodrama. *Four Saints,*
manufactured out of a musical rather than a stage tradition,
adopted its own mode as a stage work; in this it is not comparable
to similarly important works of the European mainstream, for
the latter are often enough important by virtue of their bilingual
fluency in the old language and a new one. Important American
operas, on the contrary, invent a sort of meaningful gibberish
each time out, for such tradition as was built up by the 1950s is
a perfunctory, recipe-ridden construction that genius can never
re-tailor or surmount. Unlike the operas praised in the preceding
pages, such as *L'Heure Espagnole, Der Rosenkavalier, Francesca
da Rimini,* or *The Adventures of the Vixen Sharp-Ears,* the
praiseworthy American works respond less to tradition and continuity than to the one moment—like that of *Four Saints in
Three Acts*—that they themselves create.

Porgy and Bess, on the other hand, owes something to stage tradition—to the black folk play and the emerging American musical, though nothing that had been heard before *Porgy* prepared folks for what they were to see and hear in Gershwin's opera. Where Thomson's blacks submerged themselves in a crystalline simplicity, the tonic never out of reach, Gershwin's Gullahs of Catfish Row took to the busier progressions of jazz and the uncurbed, virtually unnotatable delineation of black vocalism. Thomson's cast are blacks of mannerism, at least in the words they sing, while Gershwin's are those of naturalism. The second scene of *Porgy*'s first act, Robbins' wake, is a stunning instance of how *unlike* are the great American operas—unlike anything at all, including each other. Opening in g minor, with tonic eleventh and diminished leading tone chords alternating over a ponderous tonic pedal, the scene imposes life upon art immediately, with passionate solo keenings and a responsive choral refrain—"Where is brudder Robbins?": "He's a-gone, gone, gone, gone, gone, gone, gone"—building up to a tonal climax on an A Major triad over a C Major seventh bass, the song of the south. Sequence piles on sequence then in brighter movements as the "play" goes on in recitative and dialogue, in a hectic spiritual, "Overflow," in the widow's expansive aria, "My Man's Gone Now," and in a fanatic revivalist sermon metred like a chugging train, "Leavin' for the Promise' Lan' "—each segment leased to tension with solo or group interjections, each rabid with a racial poetry that has not since been equaled in music.

Almost nothing preceded *Porgy and Bess* and, as with *Four Saints in Three Acts,* nothing could follow it. Even the folk operas that succeeded *Porgy* bore no relation to it. Douglas Moore's longish one-act *The Devil and Daniel Webster* (1939), to Stephen Vincent Benèt's libretto, seemed to be starting all over again with another folk opera tradition, one lither than *Porgy's* and more dramatic than *Four Saints'*, with some spoken dialogue and a raft of hymns and turkey-in-the-straw. Jabez Stone of New Hampshire has sold his soul to the devil, and on the day of his reckoning Daniel Webster, "New England's pride," pleads Stone's case in trial by jury, the court handpicked by old Scratch himself; the twelve angry men are traitors of the American past and the judge is that notorious Hathorne who presided at the Salem witch trials, all called up from hell to rig the hearing in the devil's favor. "Are you content with the jury, Mr. Webster?" asks Scratch, and Webster replies, "Quite content. Though I miss General Arnold from the company." "Benedict Arnold," says the devil, "is engaged upon other business." Webster's motions are continually overruled, and the affair looks hopeless, until he addresses the jury as the men they once were in the cause of the freedom that they knew. "It was for freedom we came in the boats and ships," Webster states over a broad martial air. "The traitors in their treachery, the wise in their wisdom, the valiant in their courage—all, all have played a part." It's a naive idea, perhaps, but there is no doubting the thrill of hearing it in performance, for this may well be the only nation on earth whose open-spaced liberty can entreat even the damned. No, not the devil himself can prevail against the pride of freedom, and as the men they once were—even such vile men as they made—the jury is swayed, and finds for Stone. God knows what reward awaits them at home for that verdict.

Is some pattern or characteristic showing itself yet? Is a tradition evident? What's American in opera? Perhaps another entry from Virgil Thomson will be so kind as to inform, a second Gertrude Stein opus, *The Mother of Us All* (1947). This absurd pageant on the career of Susan B. Anthony utilized such disparate figures as Anthony Comstock, John Quincy Adams, Lillian Russell, Ulysses S. Grant, and Daniel Webster again, treating the issue of women's suffrage about as daintily as one might expect, though the question of sexual egalitarianism does charge the bulk of the event. Vastly more exoteric than *Four Saints in Three Acts* if no beginning-middle-end narrative, *The Mother of Us All* ratified its predecessor's simple modus, being precise and word-burnished, a gem, a screwball cameo. It goes here, it goes there; characters enter, argue, and exit; issues are raised and obscured; everyone and nothing happens; one can't even tell the players with a scorecard; what a treat. On the other hand, Aaron Copland's two operas, *The Second Hurricane* (1937) and *The Tender Land* (1954), strike righter postures for music drama yet haven't anything like the jaunty right-onness of Thomson and Stein. Beginnings and ends, middles and all, are not in themselves enough.

What so inspires Thomson's first two operas that failed to light Copland's is Thomson's sharp taste for the rhythm of verbal exchange, whether to the lilt of a parlor waltz or in a vocal line rapped out like a bugle call; even in Stein's frisky nonsense, what one might call the kinetic face of *The Mother of Us All* is directly read, right from the feisty opening dialogue between Susan B. Anthony and a cohort, assisted by two narrators, Gertrude S. and Virgil T. (and closed with the ubiquitous plagal tail as by a Fundamentalist's amen), right on up to the finale, which gives us Susan as a statue musing on her life on the battlements in one of the purest occurrences of C Major composition the era has known. We have learned to discredit simplicity—and admittedly there is no little guile in the reconstitution of so forgotten an art, especially in the too-too frisson of *Four Saints'* Harlem modishness—but simplicity that projects so tellingly as

Thomson's, and with such intimations of national epic about it, is not to be patronized like Doctor Johnson's dog walking on its hind legs. Ranging about the nineteenth century in decor, music, and idea, *The Mother of Us All* was planned to come off as "a perusal of a volume of old photographs"—with the understanding that they are our photographs. Listen to any scene, to any solo: there are no strangers on that stage.

Perhaps this certain informality about American opera is its most signal factor; certainly the Doctor Fausts and Stag Kings with their portentous rummage through the stuff of myth have found no place on our stages. Even from within the quaintly wrapped boxes of Gertrude Stein's euphuism, Thomson's saints and politicians speak to us with as little ceremony as Mozart's Papageno, and theirs is the carriage in which our true native tradition rides from year to year. Taking as their rhetoric the parody of psalm tunes, spirituals, ragtime, story ballads, and the what-have-you of the popular media, those American operas that carry on play in a kind of democracy of the stage, where the recesses of myth are perhaps considered too grandiose, too elite; celebrity legends, such as take in a Susan B. Anthony, a Davy Crockett, or a Baby Doe, are preferred. Failing a human legend, an animal tale will do, as witness Lukas Foss' one-act *The Jumping Frog of Calaveras County* (1950), from Mark Twain's tale of yet another Dan'l Webster, the jumpingest varmint in all Californy. Foss' gold-rushing score has little in common with *The Mother of Us All* as music, but all these very American works—sectional, even factional, as they are—bear unmistakable primogeniture in their markings, and one doesn't mean the influence of Victor Herbert or Deems Taylor; no, one doesn't. Call it the conscience of the race, if such a word as race can be at all applied; or call it the familiarity of self, the knowing the palm of one's hand.

Whatever it is, it tells, repeatedly, in terms instantly perceived or in a sly simplicity of music and word that evokes a cultural continent.

And one must be original—one must if the norm is so hopelessly normal. "Mood music for playscripts" might be a motto for much of the American operas that followed the above; going a little deeper, one might instance a fear of true comedy, of ambiguous key-relationships, of bitonality and cross-rhythms, of character, of lightning truths, of opera. Nothing is less worth reading than a list of titles, but better to list them in a clump than to spend time detailing their well-intentioned mediocrity: Hugo Weisgall's *The Tenor* (1952), after Wedekind; *The Stronger* (1952), after Strindberg; *Six Characters in Search of an Author* (1960), after Pirandello; *Athaliah* (1964), after Racine; Martin David Levy's *Stoba Komachi* (1957), after a Nō play; *Escorial* (1958), after de Ghelderode; *Mourning Becomes Electra* (1967), after O'Neill; Robert Ward's *He Who Gets Slapped* (1956), after Andryeyef; *The Crucible*, after Miller; Lee Hoiby's *Natalia Petrovna* (1964), after Turgyenyef's *A Month in the Country*; *Summer and Smoke* (1971), after Tennessee Williams; Jack Beeson's *Captain Jinks of the Horse Marines* (1975), after the onetime king of Broadway, Clyde Fitch—all based, fatally, on sources from the spoken theatre, where—considering how little these scores added to them—they belong. We have seen too many intriguing ways to adapt drama for music since *Pelléas et Mélisande* to accept these untransformed hulks. No, one must be original.

—even if only up to a point. Consider one of the more familiar American operas, technically an original, Samuel Barber's *Vanessa* (1958). Gian Carlo Menotti's libretto yields too strongly the derivation of cheap women's fiction for some tastes, but given the work's fanciful premise, it certainly does take the stage as a piece of musical theatre rather than the theatre diddled by music that one has come to expect hereabouts. Why American television has not pinched Menotti's plot for a mini-series is anybody's guess, as its soap-opera theatricality is dynamite. Eerie, voluptuous, and incisive, *Vanessa* is set "in a northern country" (a revision removed the action to upper New York state—ridiculous!) circa 1905, where folks think nothing of never speaking to their closest relatives and where Vanessa, a "lady of great beauty," has been mourning the end of a love affair with one Anatol for over twenty years. The curtain rises just as Anatol is about to arrive for a reunion—but it is his son, likewise yclept Anatol, who shows up. An amorous adventurer, Junior dallies with Vanessa's niece Erika but rides off at last with Vanessa, leaving Erika to mourn for lost love in Vanessa's place.

A perfect throwback to the days of Sardou, *Vanessa* is rich in melody, the clean, cascading lines that one had every right to hope for from the composer of *Knoxville "Summer of 1915,"* and there indeed were the tunes: Erika's "Must the winter come so soon?," the Old Doctor's waltz "Under the willow tree," and the famous quintet, "To leave, to break, to find, to keep," with its falling seventh on the last two syllables of "to weep and remember"; on the night of *Vanessa's* Met premiere one heard the humming of falling sevenths up and down Broadway as the audience scampered homewards.

Barber's only other opera was the ambitious *Antony and Cleopatra* (1966), adapted from Shakespeare by Franco Zeffirelli for the notoriously lavish opening production of the "new" Met at Lincoln Center, and more fondly recalled for Leontyne Price's getting trapped inside a miniature pyramid at the dress rehearsal than for the air of D. W. Griffith about Zeffirelli's decor; one had

the impression that elephants, not stagehands, were trunking all that exhibit into view. Somehow the opera itself failed to discharge, and Barber revised the work to smaller proportions in 1976, apparently on the assumption that the grand scale of the opera rather than the production was what did it in. True, the original did not quite encompass the whole of Shakespeare's Egyptian and Roman poetry, but Barber's first *Antony* was at least an opera; his second version consists of scenes from an opera, a concert that affects decor and staging as a subdebutante affects wordly argot. As with *Cardillac, Der König Hirsch,* and *The Bassarids,* composers should beware advice.

In any case, Barber's romantic route remains the one most often taken by successful practitioners of American opera composition. Utilizing a distinctly national sound, Carlisle Floyd enriched the repertory with *Susannah* (1955), set in Southern hill country, where the heroine is too distinctive a figure to blend in well with her intolerant neighbors—which is more or less the centrus of Floyd's most recent work, *Bilby's Doll* (1976), though here the nonconformist actually thinks she's a consort of the demon world. Like many another contemporary composer—Britten, especially, comes to mind—Floyd deals in the touchy relationship between the lone-wolf iconoclast and the rest of society; this is the focus of theatre and literature today, and so it must be, Johnny-come-lately, with opera. Floyd's short monodrama, *Flower and Hawk* (1972), looks in on Eleanor of Aquitaine's durance vile in Salisbury Tower, his adaptation from the novel *Wuthering Heights* (1958), has as its principals Emily Brontë's untamed moor-beasts Heathcliff and Catherine, and his one international success, *Of Mice and Men* (1965)—adapted from John Steinbeck's script rather than his novel, which was rather scripty to begin with—likewise details the misadventures of misfits. Yet the composer himself paints these strangers in unstrange music, as if his audience can accept the world of the outsider only within the context of the familiar, the tonal, the reassuringly melodic. *Wozzeck,* now—this is the sound of the outsider. But how far afield can our rigorously democratic art voyage? Like some others before him, Floyd troubled to write a preface, anent *Susannah,* on the nature of the composition of words and music in opera, addressing the difficulty of instilling the sacred communion in a culture perhaps too amply endowed with a predilection for the profane, "a necklace of arias, duets, and ensembles strung together with brief jarring recitatives" in Floyd's classical model. "To expect absorption and belief in what is going on on the stage under these circumstances seems to me to be blindly and absurdly optimistic." This is a paradox of American life, and a dilemma for American art: en masse, the culture fears nonconformity, yet seeks always to individualize itself in heroes, stars, mavericks . . . dissenters.

Not surprisingly, one of the most popular American operas, Douglas Moore's *The Ballad of Baby Doe* (1956) is also devoted to the iconoclastic road not to be taken. John Latouche's libretto, based on chronicle, relates in operatic fabric a subject fit for true ballad: how Baby Doe, "the miners' sweetheart," maneuvered her way into wealthy Horace Tabor's life, fell in love with him despite the difference in their ages and the fact that Tabor already was married, threw her lot in with populist silver for Tabor's sake when gold was inexorably clasping the standard, and, sworn to hold the Matchless Mine that gave Tabor his first break, was eventually found there frozen to death.

There is no narrator in this "ballad" of Baby Doe, but all that's lacking is a diatonic refrain for familiarity and the episodic strophes for storytelling. This is true national opera, as conserva-

tive as all get-out until the final scene, when a grotesque narrative flashback and -forward, borrowed from musical-comedy technique, puts a little juice into format. Flair for era pervades both
music and scenario—President Chester A. Arthur and William
Jennings Bryan both turn up on stage, and ragtime, waltz, and
ballad proliferate—and while there are tunes to recall, Moore is
at his best in the vocal passages between, that space of expedients
where the recitative used to dwell. The most distinctive such moment is the end of a scene between Tabor's castoff wife and her
vindictive friends, who recommend exposure of Baby Doe and
condemnation of Tabor: "Shout it from the housetops!," they
shriek three times, in metre exactly paced on the spoken cadence
of the phrase, and then, for good measure, a final "housetops!"

The national success of *The Ballad of Baby Doe* does not
prove the valence of a conservative tonal idiom, but rather the
attraction of effective musical theatre in whatever idiom. Since
Moore's death, the foremost composers of American opera have
been Floyd and Thomas Pasatieri, but even Floyd's *Susannah*
and *Of Mice and Men* and Pasatieri's *The Seagull* (1974), from
Chekhof's play, do not address the public with such panache as
do Moore and Latouche in their cut of Americana—and this is to
leave out altogether the expressly theatrical play-operas so favored in the fifties and sixties. *Susannah*, of course, is a landmark
of regional art, and *The Seagull* is one drama that works well as
opera, but *Baby Doe* is original in concept and that makes all the
difference, in its eleven short scenes, each carefully end-stopped
(rather boringly so, in fact), its abundance of spotlight solos running a thin line between authentic folk parody and musical comedy, its breathless, almost rabid temperature, pushing early in
the evening into the obsession with doomed silver that almost
wiped the west off the political map for a generation. Latouche

passed up the chance to manipulate adultery and silver into symbols, and thus less into more, but there is even so a preceptive sophistication in his outwardly simple text, and a farouche determination in his heroine, as when she hands her jewels to Tabor, the very picture of American dissent: "I've always loved a gamble," she cries, and adds, climbing up to a B Flat, "Place my bet on silver!"

Latouche's libretto peaks in the last fifteen minutes, when the broken Tabor revisits the deserted opera house he once owned and sees, replayed in slithery fantasy, the events of his life, both conscious and otherwise. Here the folk ballad cedes to dramatic initiative, and while the psychological intermezzo is hardly a novelty in theatre, it is just that in American opera. In Tabor's dream, his rise from New England stonecutter via marrying the boss' ugly daughter to Colorado miner and plutocrat is enacted almost cinematically until the vision turns bitter and its tentacles disperse as Baby enters to comfort him with an aria that sounds like the issue of a hymnal and a pianola. In its trio section, Baby puts back the hood she has been wearing, revealing her hair—white, with age, for much time is now to have passed—and as her song draws to a close, the Matchless Mine appears, her tomb, and snow begins to fall.

All that Moore's opera needs is an imaginative modern staging by a major company; its immense vitality is at present more proposed than experienced, for American opera production is hampered by a paucity of first-rate and even adequate stage directors, though the present younger generation of singers can do anything composers, librettists, and the great European producers ask of them. If exacting music drama is the weakness of American opera, it is also the weakness of most American opera companies, which would be sorely pressed if they ever attempted to mount a production of such few truly avant-garde pieces as have originated in the U.S. Boston hosted the stateside debut of Roger Sessions' atonal *Montezuma* (1964) in 1976, much to the dismay of Bostonians, and the Juilliard School of Music did a smashing job on Harold Farberman's *The Losers* (1971)—two exemplars of America's slim output of antinational, Continentally influenced "modern music." *Montezuma* is the larger piece, with a heavily literary libretto by G. A. Borgese and the epic distance of narration and historical conflux of conquistadors and Aztecs, while *The Losers* squats in the present-day America of the motorcycle gang, hip and slangy in Barbara Fried's text, and generally less vocal than Sessions' score, which the composer requested be sung with no little bel canto. Farberman favors the tone cluster; rather than describe the contours of the action, he gives the whole a general atmosphere, punctuating the idiopathic subculture of the hoodlums rather than the rise and fall of Fried's lines. Neither work is likely to find a place in the local continuity, and even these single engagements point up the gap between what is acceptable and what isn't in America, where art, like any other business, must pull its weight in the market and reassure the public of its democratic applicability—useful to all, like a subway turnstile.

Some attempt has been made of late to harmonize new works with old audiences by welcoming "theatre" directors into the opera house (as if opera weren't "theatre"), but still the material itself must tell or the experience is nothing. Even Tom O'Horgan, known for peppering up tepid scripts with pictures and capers, could make nothing of the New York City Opera premiere of Leon Kirchner's *Lily* (1977), an incomprehensible debauch of Saul Bellow's novel *Henderson the Rain King*. Composing music of no dramatic feasibility—of no shape or presence—and writ-

ing his own libretto, Kirchner covered only a quarter of the original and built up the role of the hero's wife out of all proportion;
the result was exactly comparable to an operatic version of *Hamlet* that ends with "The Murder of Gonzago." Seldom has so
little added up to so much less; here was a prime example of the
new work that despoils both the tradition and the future of opera
at once . . . a ripoff, literally. The American operagoer is conservative enough as it is without having a *Lily* to point to and
cry "aha" about.

So conservative is the operagoing community, in fact, that it
was unable to respond even to Virgil Thomson's latest opera,
Lord Byron (1972). One could hardly call Thomson's musical
signature "difficult," even this late in his career, but still it proved
too singular for most ears, too dapper in its own idea of mode—
and this despite the clear narrative approach of Jack Larson's libretto, nothing like the gaming enigmas of the two Stein operas.
Byron, at curtain rise, has died; the bulk of the piece dramatizes
his colorful memoirs, the disposition of which the late poet's
friends are debating: publish or let perish? As spry as ever, the
vieillard terrible, Thomson has his joke in that, although they
ultimately destroy the manuscript, we the audience have seen its
secrets revealed anyway. Thomson played other jokes as well in
Lord Byron, exercising his taste for neoclassical parody, burlesque, and quotation; his mobilization of form is a seminar in
opera construction, certain, one day, to be assimilated—provided
the egregious suction of foreign patronizers and the soi-disant
"critics" of the New York media is ended.

Profane Thomson, the hieratic madcap, brings us back to this
chapter's inaugural postulate, that America's opera is essentially
both profane and popular—thus its noted informality—closer to
comédie mêlée d'ariettes, ballad opera, and *Zeitoper* than to op-

era, but occasionally exploiting the musical facilities of European "opera" for romantic power and thereby cutting across the barrier of genre classification. In Europe, operas are seldom produced in commercial theatres for an open run, but several of Menotti's works played successful engagements on Broadway, and in this adaptability of the popular theatre lies the weakness of the American opera tradition but the strength of American musical theatre. A prototype of the profane—naturalistic, parodistic, encouraging disbelief—finds its likeness on Broadway, where numerous trained musicians have helped create a form that at moments seems more musical than some of the operas that have turned up for title in the preceding pages. Kurt Weill, for example, wrote three one-act operas for Moss Hart's play *Lady in the Dark* (1941), each a through-composed dream sequence in the mind of the troubled heroine, recitative, ensemble, and . . . aria? No, but who can draw lines with any confidence now that the tenets of the ultramusical romantic era have been overturned by the judicial review of the neoclassical court? After passing seventy years as a spurious opera, a demimondaine of the sacred world, Bizet's *Carmen* was finally revived as the *opéra comique* it was intended to be, on Broadway, as *Carmen Jones* (1943)—yes, in English translation, reset in the American south and Chicago, and in a reduced orchestration, but *Carmen* all the same.

The originality that has largely failed American opera has driven composers to place their Baby Does in a more overtly theatrical atmosphere, where they are assured of the chance to reach a public with the clarity that one demands of the commercial theatre. For every Dominick Argento, who with a work such as *Postcard from Morocco* (1971) uses tropes of the spoken theatre (seven puppet figures monologuing "in Morocco or some place"), there are five or six Argentos working what theatre folks refer to as the Street to reach a less penetratingly musical but insistently word-oriented public in the commercial theatre. Marc Blitzstein turned Lillian Hellman's play *The Little Foxes* into *Regina* (1949) for the Street, though it has since wound up in opera houses; it should have stood as an object lesson in how to convert playscripts into librettos before the music is applied, for there are no more *Pelléas et Mélisandes* to be written, however often American musicians attempt to write one. Blitzstein, as always his own librettist, reevolved Hellman's structure to tell her story in her language, yet so elastically as to allow for musical contiguities not implicit in *The Little Foxes*. An indictment of capitalist greed as viewed in the mercantile and personal complots of a corrupt southern family, *Regina* benefits from the word-clearness of the profane theatre, adapting the *recitativo stromentato* of the Mascagni-Puccini era to a more fluid dramatic course; it is weak in the quality of its music, but not in the circuits through which that music is run. Blitzstein was not the world's greatest librettist, either, and some of his "operatic" constructions on Hellman do not come off, but, proving the theatricality of the vernacular theatre, he created a splendid curtain for Act II: at a gala party, the evil heroine Regina taunts her husband Horace with his ill-health and probably imminent death while their guests enjoy a gallop. "I hope you die," she tells him, unheard amidst the noisy dancing. "I'll be waiting for you to die!" But when the dancing comes to a halt, the orchestra ceases, and Regina, lost in her harangue, screams "I'll be waiting!" on a top C. Shocked, the guests turn to stare at her. Silence. The curtain falls.

If the tensile, maverick punch of the profane is a hallmark of "Broadway" opera—a classification that would extend to a number of European works as well, from *Die Zauberflöte* to *Opéra d'Aran*, neither does it lack for the contraposition of neoclassical parody. Jerome Moross and John Latouche, they of the *Ballet Ballads* in the chapter on national opera, petitioned the theory of *Gesamtkunstwerk* for one of the most refreshingly native entries in the repertory, *The Golden Apple* (1954), unlike *Regina* an event totally through-composed and utilizing a formidable battery of pastiche numbers in the vernacular idiom for a kind of polemical vaudeville. Latouche's libretto updated Homer to early twentieth-century America in order to treat the lure of economic power in a nation just beginning to prefer technocratic imperatives to its priorities of earth-salty wisdom . . . really, what resilience is tendered in this form, what a way to justify the Word to the cause of music, what a *comédie* is here *mêlée d'ariettes!*

But it is just that mélange that disputes the genre of such works, that forces *Regina* and *The Golden Apple* into the twilight in books such as these. How often before we've spotted some work or other pure in its profaneness—Roussel's *Le Testament de la Tante Caroline,* or Weill's *Die Dreigroschenoper,* for example. But they were exceptions, subordinate clauses in the world's debate; here we are faced with a whole tradition, and one, confusingly enough, that does not stint itself in musical expression, in a nonpopular ambiguity, in such guises as Blitzstein's Brechtian exposé *The Cradle Will Rock* (1937), Weill's urban melodrama *Street Scene* (1947), or Leonard Bernstein's satiric fantasy *Candide* (1956), the last yet another victim of disastrous revision.

Candide in the current scaled-down version of 1973 is a trivial frolic scarcely worthy of mention, but the original adaptation of Voltaire's blast at Enlightenment optimism offers a telling example of how the juggling of the sacred and profane may confuse rather than enrich point of view. Not as clear as they might have been on whether they were waxing romantic or satiric, Bernstein and his wordsmiths Richard Wilbur and Lillian Hellman (with some help from John Latouche and Dorothy Parker) veered bafflingly from the self-commentative profane to the self-absorbed sacred and back again, muddling what otherwise remains one of the masterpieces of its kind. A "comic operetta," they dubbed it—and so, in the main, it was, with a smashing rondo overture (complete with Rossinian crescendo), an extended spoof of that curious genre, the jewel aria, for Cunegonde, "Glitter and Be Gay," a wicked paean to syphilis for the Leibnizian Dr. Pangloss, *Allegro con spirito all'Ungarese*, "Dear Boy" (not, ultimately, used, though awaiting restoration in the published score), a flowing chorus of pilgrims undermined by major sevenths and ninths in the Alleluias, a baldly profane spoken interruption of the first-act quartet finale by an acid Voltairean, and so on. Yet, conversely, there were two moving folk song solos for Candide, his lovely ballad in $\frac{5}{8}$, "Eldorado," and an eloquent finale, "Make Our Garden Grow," built to titanic effect out of Voltaire's unassuming final line, "Il faut cultiver notre jardin."

Certainly, the profane far outweighed the sacred in Bernstein's *Candide,* not least because the work belongs to the distinctly profane phylum of the Broadway musical—but then Weill's *Street Scene* is of the same order, and it almost never uses satiric leverage in addressing the spectator. Perhaps one should underline Weill's *Lost in the Stars* (1949), one of the more sacred offerings of its race (what to call it?—an opera with words?) and a far cry from the reprobate *Verfremdung* of *Die Dreigroschenoper*. With a libretto by Maxwell Anderson drawn from Alan Paton's novel, *Cry, the Beloved Country, Lost in the Stars* borrowed Paton's South Africa to continue the history of black American folk opera, for Weill had so naturalized himself by then that the jazz of Johannesburg speaks of hot Harlem and a tender song about the "little grey house" at home in Indotschen has the rolling bass of many an American country ballad on a similar theme. In truth, Weill did much more for American music in this African piece than he did in the melodramatic one-act, *Down In the Valley* (1948), a tale of true love thwarted that borrows its tunes, not all that triumphantly, from the folkish panorama.

So far we have come from the days of *La Buona Figluola* and *Die Zauberflöte* in material, yet not so far in form. Comedy, that gay dog, has not changed—how could it possibly?—but opera is very much altered, for comedy does not bend the way music does: comedy is our brittle messenger, but music is our consolation. Now that we have seen the two make their peace, we must move on to the last of our three pivotal chapters. The first taught the sacred acknowledging the profane; the second, the emphasis of the profane and its reclamation of the sacred. The third, now, celebrates the modern treaty of the two. This one belongs to Godot.

VII

Three Scarecrows
and a Pair of Feet

Simply put, the art of modern opera is a syzygy of the sacred and the profane—more exactly a "profaning" of the sacred—a development that can be traced directly from the textual clarity of *Pelléas et Mélisande* through the neoclassified romanticism of *Wozzeck* and the contraposition of *Jonny Spielt Auf* to the right-angled intermezzo of *The Bassarids*. What in Richard Strauss' day could be broken up into words and music for the purposes of disputation has become, for the first time in opera's history, a synaptic organism, one in which the musical entity of romantic opera and its pull to fantasy is recycled against a simultaneous diagram of comic commentary—the horizontal extensions of song tempered by the vertical analysis of the Word.

No longer strictly buffo, melodrama, pastoral, or what-have-you, opera is bimodal in the age of Godot, when satire closes with romance and tragedy and takes them over. Interestingly, the near-approaches to this veritably modern aesthetic are centuries old. One recalls the juxtaposition of nobility with naturalism in Busenello's *L'Incoronazione di Poppea* and, more crucially, the influence of a more aggressive musicality than opera had been used to in Monteverdi's *L'Incoronazione di Poppea;* and of course there are Mozart and Da Ponte and Mozart and Schikaneder, the latter exploiting the nobility/naturalism of that early double-genre, the rescue opera. But these were affiliations, not consummations, matters of temperament more than form. In modern opera, the two poles of musico-dramatic ploy do not trade off but combine, as they have been mixed and remixed throughout the twentieth century. From mannerism to expressionism to repurified neoclassicism, opera has retrieved its ceremonial myth in audaciously anticeremonial ways, but the gambit of comedy, whether gamed as alienative parody, intrusive intermezzo, spoken dialogue, or funhouse spoof, has not lightened but deepened music drama, put grain to its images as one bevels the glass of a mirror.

Moreover, the continental dissolution of national borders has emancipated a form more strictly attentive to individual drama than ever before, using the ethical language of the classicist to ratify the urgency of the romantic; *Wozzeck* would be the clearest such example as a pivotal point. Its children are uncountable, as are those of *Duke Bluebeard's Castle,* a pivotal point in the opera of psychological myth and further evidential of the continuing focus on the solo protagonist and his crisis of identity. That focus we toast herein.

Five operas, now, to view before the curtain falls, singular works to prove the pivots—one moral comedy, one mythic farce, one epic, one absurdity, and one satiric tragedy. One had fears that a volume on twentieth-century opera would fade to a feeble close: who would follow Strauss and Britten without anticlimax? *Coraggio.* Lashed, but not leashed, to technique, come wonders. "Piously I cherish the old," avers La Roche in *Capriccio,* "patiently awaiting the fruitful new, watching for the masterworks of our age!"

Igor Stravinsky's *The Rake's Progress* (1951) is a test case in neoclassical conceit. Using Hogarth's renowned picture-cycle as a starting point, W. H. Auden and Chester Kallman wrote a three-act libretto, mainly in verse; Stravinsky's source was more comprehensive, though in the main relating to Mozartean *dramma giocoso.* Neoclassicism made for the public as much as for the scholar, *The Rake's Progress* is an etude quickly studied and, despite its aridity of ostinato, easy to love. The score is recapitulatory in ways not known to Mozart, and of an aural intelligence, formalistic so that the ears may "see" at first hearing, even if the concertante of aria and ensemble is inventively modern in reasoning and concise rather than simple.

For "Mozartean" music, a plot for the eighteenth century. Auden and Kallman plucked their characters from the registry of Restoration comedy—Trulove, Rakewell, Nick Shadow—to record the rise and fall of a lazy sensualist, loved but not loving well, whose three wishes (for money, happiness, and, alas, too late, love) are seen to by the devil, and who profits from such illegitimate assistance by ending in Bedlam. Tom Rakewell, the tenor of the evening, goes off to London for self-expression on an infernal subsidy, and in the city he learns the catechism of the sweet life in Mother Goose's brothel, marries the town topic, a bearded lady, to demonstrate the nullity of responsibility, and then faces the reckoning that catches up with all such rakes in a game of cards with his benefactor. Not surprisingly, the devil cheats, but true love in the voice of his sweetheart Anne intercedes for Tom and he wins his freedom—for which subversion of practice the evil genius decrees him mad, and of madness in the madhouse Tom dies, succored but, again alas, deserted by his Venus. Is that sufficient? No! Out before the curtain the principals pour for an epilogue, minus wigs and appurtenances, as the house lights come up. Says the town topic, Baba the Turk: "Good or bad, all men are mad; all they say or do is theatre." Says Nick Shadow: "Many insist I do not exist. At times I wish I didn't." And say they all in concert: "A work, dear Sir, fair Madam, for you and you." Bow and exeunt.

No selection of excerpts does justice to Auden and Kallman's text, one of the most literary of its kind, and as for Stravinsky's music, it comes near to redeeming the age. How grandly the pilot of the storybook *Nightingale,* the shrilly grave *Oedipus Rex,* and the buffoonish *Renard* brought his ship-of-the-line to port, and how elegantly his shapes delivered the verses, rhymes and all. Is this the neoclassicism we reviewed at the midpoint, when *Wozzeck* came? No: *Wozzeck*—and *Lulu*—sought the core of romantic naturalism in the musical dialect of *logos; The Rake's Progress* is unalloyed classicism, respectful of the world code, laid out in the symmetry of the discrete forms for which Berg offers no counterpart. His forms are row variations and whole scenes; Stravinsky's are repetition and episodes. *Lulu,* the ultimate of Berg's art, is algebraic in form, whereas *The Rake's Progress* is anatomic, for while Stravinsky reveals the structures, serialism deals in infinitesimal multiplicity of substructures.

This doesn't even take in the question of tonality, relevant to which *Lulu* is naturally the more difficult work by far. Furthermore, each of this pair was born of an era different in hope and attainment from the other. *Lulu,* from the late twenties and early thirties, was conceived when serialism was the strange device by which romanticism could at last be set free on its own terms, its passion no longer engulfing its speech, while *The Rake's Progress* is a product of the post-World War II reconstruction, when formalism could no longer call itself the experiment. On the surface, their two plots deal in the same adventure of rise and fall, and much is contained in the two composers' different methods—in *Lulu's* realism, *The Rake's* mannerism; in *Lulu's* tragedy, *The Rake's* farce; in *Lulu's* grimacing orchestra, *The Rake's* courtly consort. But couldn't *Lulu* have been written later and *The Rake* earlier? The point is that they weren't, that these two exceptional works prove rules of their time on the subject of dramatic music, rules not meant for application by anyone else.

One-of-a-kind shots, summations of theorem apparently antithetical but having *logos* in common, the romantically profane *Lulu* and the neoclassically profane *Rake* are at any rate not exceptional in that they are after all operas and as such have much in common with *La Traviata* or *Die Fliegende Holländer*. True, *The Rake*'s opening fanfare, like *Lulu*'s prologue of animal crainer and menagerie, strikes a distinctive tone. No sooner has Anne given the first line, "The woods are green and bird and beast at play . . ." than one knows how unlike anything else one has heard this work will be. Like, yet unlike—that seems to be the course we've been running these fifty years:

> Now is the season when the Cyprian Queen
> With genial charm translates our mortal scene,
> When swains their nymphs in fervent arms enfold
> And with a kiss restore the Age of Gold,

sings Tom to Anne, describing more than being. But he can be, too, for after turning down the offer of a job from Anne's father, he reasons that fortune, not merit, calls the steps of the dance, and anticipates his progress in a lively, extroverted air, being for sure. "Come, wishes, be horses," he declares, "This beggar shall ride!," holding out the last note while the orchestra canters gracefully under and then, four times, neighs.

Period, manner, poetry, and all, *The Rake's Progress* is still a realistic piece behind the fantasy, skirting awareness at each thwack of the harpsichord in recitative; Hans Werner Henze's *Der König Hirsch* (1956) shows how antirealistic modern opera can be, reaching for myth, reaching for the moon and its looted emanation. Henze's librettist Heinz von Cramer looted Carlo Gozzi for this *King-Stag* (Gozzi himself had made free of a pre-Hellenic antlered-quarry hunt such as crops up continually in Celtic legend), emphasizing that parity of romance and knockabout comedy that Busoni and to some extent Puccini disinterred in their Gozzi operas. Through von Cramer's central storyline, on the self-realization trauma of a king who tries to escape into the peopleless green world of the wood but ultimately returns to rule his subjects, runs a congeries of contraposition, parody, and ballad-comedy high-jinks not unlike those of *Die Zauberflöte*; in a gang of improvisatory clowns, called "the alchemists," is found a wellspring of rude mechanical slapstick and modern alienative technique.

One of the longest operas of the day, possibly the longest, *Der König Hirsch* packed so much action, dialogue, and polyphonic orchestration into its three acts that Henze could dedicate each third separately to different people without looking stingy. Yet the entity, like Wagner's longer works, depends on the romantic's cyclic unity for thematic concentration, even amidst the neoclassical structuring. With so many characters—not only the king, his Governor, a maiden, the clowns, and two comedians, one a moony peasant lad and the other a "shy assassin," but assorted personages from the land of enchantment roam freely in and out of the action—a certain jumpiness is unavoidable, but if this is Wagnerian romance, it is cut with the tart bustle of Shakespearean comedy. Indeed, the spirit of *A Midsummer Night's Dream* haunts the proceedings as much as does that of *Die Zauberflöte*, not least in the climactic metamorphosis of King-into-Stag (though unlike Bottom, Henze's King is no comic, but a Wordsworthian hero seeking the inner passages of nature in escape from his urban realm).

"Who comes out of the forest avoids no boundary; Who goes among man learns fear" is the somewhat opaque advice given the King by two statues (women, not armed men; still, Mozart is not far behind). Yet the maze of mythical orientation has its ambiguities after all, and the puzzle is one of its perquisites. King or Stag, ways lead both in and out in all directions—into the wondrous forest, into the city, a "Venice between wood and sea"—into the transformation of personality and the identification of good and evil. In the elaborate finale of Act II (later removed, the voice parts orchestrated, as Henze's Fourth Symphony), the King confronts the forest, spoken for by an unseen mixed quintet, and demands the right to discover. The height of romantic rhapsody, an encounter out of nineteenth-century poetry, the scene gathers up the strands of serialism and rondo form as if to subdue the complexity of program with the assurance of system. "Open and let me in!" cries the King. "You don't belong to us!" the forest answers. "Give us peace and go." Restless and urgent, the orchestra pleads with the King, *espressivo, cantando, appassionato*—an Italian opera, this, of sorts, Italian in source and voice, though Nordic in complexity.

In fact, *Der König Hirsch* is perhaps too complex a work for its own good, too heavily scored and too rampant with its puzzles to deliver recognition and catharsis to the public's satisfaction. But if the opera's romantic shaft dives into mystery, its quotient of the profane is, naturally enough, clear to view. At the finish of Act I, when the King's dastardly Governor hands his bashful henchman Coltellino a pistol and tells him to follow the King and kill him—when the action has achieved its first minor zenith —enter the clowns on all sides, without any warning, to warm legato string chordings against a brainless drum and cymbal parade, circus music lacking the music. "A gala!" cry the clowns. "A grand and special gala!" They have come to perform for the King, and are not indisposed to perform for themselves, so Henze directs that their play and byplay be, as it was in Gozzi's day, left to the spontaneous combustion of the clown team, who do not sing but speak in metre and bang on wood and metal when not improvising their drollery. In their element—they're on stage, are they not? that is life to the clown—they resort to an ear-busting serenade, reminiscent of the work of Harpo Marx, which brings the Governor out. He dismisses them, saying that the King has gone back to the wood, and as the downcast clowns repeat his words, two echoes repeat theirs and a hush comes over the action. The woodwinds sound a minor triad, the strings borrow it, shuddering, and a plangent solo viola closes the act to a slow curtain: back to the sacred.

This constant engagement of and detachment from the commedia in *Der König Hirsch* is, in a way, the new myth of the modern theatre—that of comic naturalism, though its voyage is a short one, its catalogue undistinguished, and its descent into the underworld something of a Hallowe'en skit. It is up to the sacred to find a way into or out of the wood, and ultimately the King materializes bearing the antlers of a stag to face down his marplot of a Governor. With the minions of fantasy protecting the King, evil cannot win: wind spirits shield the King-Stag from assassination, and a parrot pecks Coltellino's shoulder, causing the gun to go off and kill the Governor. Coltellino is thrilled, for he has always been shy at his calling, and the King quotes from the message given him by the statues: "What transforms, can assist; it helps, what threatens. What does not kill, transforms; it helps, what threatens."

With all that occurs in *Der König Hirsch*, it is hard to tell whether or not its puzzle is fully solved in fine, especially since the 1962 revision, *Il Re Cervò*, unfastened the romance and featured the high-jinks, covering the holes left by an hour's worth of deletions through the use of an alienative narrator. But whether or not von Cramer's text proves its hypothesis, *Der König Hirsch* is vital art in form, all the richer for its juxtaposition of the cruel and the crazy with the metastasis of romance.

Opera is in the main a matter of instinct, however well Gluck could analyze his contrapositional violas, however much background reading Wagner did before undertaking the *Ring,* and no matter how clearly the critic thinks he sees. There is much to draw on the Western unconscious that isn't necessarily "known," and once the rudiments of counterpoint are digested, genius gets it right by sensibility far more than by plan. Henry James' "associational process" connects not only the gut reactions of the spectator to the work, but the gut creativity of the *author* to the work in the first place. Perhaps Henze and von Cramer intended some denotation in calling their clowns "alchemists," perhaps not —but the satiric mentality that these clowns embody is truly the alchemy of modern art, converting romantic idealism and classical rationality into the aleatory *logos* of Godot. Parody and the intellectual stab of the double-bar line are everywhere now, whether in the Hogarth and Mozart of *The Rake's Progress,* in the qualified romance of *Der König Hirsch,* or in the epic contraposition of Bernd Alois Zimmermann's *Die Soldaten,* next on the bill. Such dissimilar forms as these three have one point of reference—era, the times and its mores; and the times call for the *gai primitif* in music as well as Word. Henze speaks to Tippett in Act II of *Der König Hirsch* in a canzone, "Als ich mit Vogelzungen noch," accompanied by guitar—but unlike the nostalgic atmosphere Achilles spells in *King Priam,* Henze's canzone is a calmer, almost reasonable piece, though addressed to the four winds in standard ballad stance. No matter: Henze speaks to Tippett, and Tippett speaks it onward, to others. The toss, master to master, is that of method, as it always was: of form.

Dissimilar is certainly the word for the form of *Die Soldaten*

(*The Soldiers,* 1965), an undertaking of mammoth technical menace originally designed to be played on twelve separate stages on the perimeter of a circle surrounding a swivel-chaired audience, who would have to keep rolling eyes on all four points of the compass even to catch the dimmest idea of what was going on. The "ballshape [*Kugelgestalt*] of time" was Zimmermann's stated aesthetic for *Die Soldaten,* and his pluralistic arena was planned to capture a simultaneity of past and future so as to render the present as a morsel of a whole—again, the moment, not the scene, of serial opera, the direct statements, one, two, three, four. Zimmermann's idiom is very modern, but he hammers out his kinky time-scheme in music as well as action by quoting old masters and new masters, all as part of his startless, finishless, middleless globe of time passage.

What other opera does this recall? Yes, *Wozzeck;* Berg's *Untermensch* has been a rallying point for many of the works between then and now, and *Die Soldaten* particularly shows Bergian colors in its source. Taking as his text Jakob Michael Reinhold Lenz' strange play of 1776, an early romantic entry on the bestial vanity of the Prussian military caste, Zimmermann found a sympathetic collaboration for his *Kugelgestalt der Zeit* two hundred years before the fact. In Lenz' bizarre scene plot, a myriad of large, medium, small, and tiny sections of time (thus the connection with Büchner), the soldiers of the title run roughshod over everything in their path like an army on the march, utterly immoral though respective of a sort of class ethic, and Zimmermann caught Lenz' velocity so well that this overloud and difficult score has conquered conservatives as well as progressives. From the tyrannical drumbeat of the prelude to the thoughtless tread of the soldiers at the finale, *Die Soldaten* does what few serial stage works have been able to—express. But music, said

Debussy, is made for the inexpressible. Zimmermann's inexpressible is that of Schoenberg and Berg—that is, the disheveled romanticism of the underground that for all its method is rich in the madness of drama. Zimmermann's particular madness is not unfriendly to the expressionist vocal line of the old avant garde; unlike many of his contemporaries, however, Zimmermann is always making music. It was perhaps farfetched of critics to compare *Die Soldaten's* third-act female trio to a similar picture in *Der Rosenkavalier*, for there is nothing similar about them— but the very suggestion shows how potent this modern vocalism can be.

Zimmermann's multifarious time form with its twelve stages and swivel chairs proved too remote a proposition for the logistics of opera companies as they stand today, so he compromised with the proscenium, imposing a multitiered stage and a projection screen upon it. No doubt *Die Soldaten* offers a genuine arrival of the "stage action" of the modern era, liberal politics included, for Zimmermann lets the irony of double- and triple-action expand Lenz' already fierce indictment of the war machine so taken for granted in the eighteenth century and so ripe for indictment today. No mere revival of Lenz's play could possibly compel as does Zimmermann's realization of it, with its osmotic barrage of past and future to inform the present indicative. For example, the crucial scene in which Baron Desportes seduces the heroine, Marie Wesener, is played against two other scenes, all interlocked. We see Marie write to her suitor Stolzius; we see Stolzius being comforted by his mother upon receiving the discouraging letter; in another part of the forest, we see Marie's grandmother in her wheelchair grieving over Marie's imminent downfall—and meanwhile Desportes has arrived on the scene and that downfall, from burgher's daughter to camp whore, is undertaken by the unscrupulous officer: three scenes on one idea, one scene in three, solos, duets, quintet . . . the transmission amplified by the by-now celebrated burst of coloratura hysterics from Marie, the damage commenced with a fireworks display.

They thought of *Wozzeck,* writing about *Die Soldaten,* searching for references—as did we—for both Berg and Zimmermann rescued old novelties of drama to turn them into new novelties of opera, to focus on the individuals swamped, as the romantics saw it, by the status quo. Individualism again the thesis, the system again the antagonism—this is the ism of European opera that replaces national artwork, forcing the ear to accept the statement rather than the chords. Despite the generalized appeal of Wozzeck's "Wir arme Leut!," his reminder that "Man hat auch sein Fleisch und Blut!" exerts a stronger response, and both *Wozzeck* and *Die Soldaten* are their people more than their ideas. As even the generalizing Hans Werner Henze was driven to tell the critics who accused him of pantywaisting the progressivism of art, operas are written by individuals, not movements. Much as one may trace trials of mannerist romanticism from a theatre of tunes to a theatre of word tunes—from *melos* to melodrama— these are people one speaks of, individuals, not movements. The form, more than ever now, yields a deeply personal character; the system is disallowed.

One of the few postwar works not by an established master to be granted multiple productions, *Die Soldaten* reminds us that the intriguing works that must fund the repertory of the future are coming still and are yet to come. New operas are exciting; the synergic resources of the world-wise and the balladeer signal vitality sure, and given modern stage technique, there are wonders to be worked truly unexpected in the encounter as well as tutelary in design. Those nos that battled the yesses at the Berlin premiere of *Der König Hirsch* (oh, the many minutes of clapping and pfui, the arguments in the coffee houses, the reflection in the press—they still speak of it as a great moment in the opening of the operatic forum) are giving way to qualified maybes, and such formerly conservative territory as England can take its *Punch and Judy* along with its *Peter Grimes*. Once they asked, and asked again, where is the next Strauss, the next Massenet? But even Massenet himself was no longer the first Massenet at the end, and the Strauss they adored had advanced into neoclassicism, leaving sensuality to sit below the salt at his table. Instead, there will be others, newcomers. The catch, if transformation can be called a catch, is that such as Zimmermann await at the turnings in the route: Massenet and Strauss and the way we were are dead. You wanted *Louise* before, and *Louise* was yours for the asking. But you got *Pelléas et Mélisande,* too. That is what time entails. Whether the source be Lenz' or Gozzi's plays of two centuries back, or a modern drama, as with Iain Hamilton's *The Royal Hunt of the Sun* (1977), or the morale of epoch, as with *The Rake's Progress,* or no source at all, this an era to enjoy now, with all its faute de mieux, excess, miracle, trackless probe, and report.

Take, for instance, Stanley Silverman's *Elephant Steps* (1968), a "fearful radio show" based on nothing, literally. Pastiche counts so heavily for Silverman that one is never certain when he is guying and when sincere, and so artless is the "artificiality" that ultimately all of Silverman is sincere, and all of him guys. Madrigal, gypsy tango, electronic tape, serialism with the idiot celesta and xylophone solos, as pregnant as a summit conference, late vocal ragtime, vaudeville "traveling music," the meticulous twang of Indian raga, and moderately heavy-metal rock put over a jobbery on the public, a very pleasant jobbery that is no less fundamental in its address of the national whereabouts than Thomson or Moore are in their ambit.

Why all the glitz? What, if anything, lies below the eclectic homage? It's hard to tell for sure, as Richard Foreman's libretto is as abstruse as Silverman's omnium-gatherum is easy to place. Apparently this *Elephant Steps* is about Hartman, an ailing visionary who, in his sickbed, is given a deal of useless advice by other characters, makes his way to a radio station where he is urged to expose "the evil ones" (Hartman doesn't), then prowls the streets at night seeking his guru, Reinhardt. Strange, not to say unearthly, characters try to befuddle him (they can't) and, climbing a ladder, Hartman sees Reinhardt and in an ecstatic crisis believes himself to *be* Reinhardt. Through it all, elephants sneak around to, if no purpose, much purport.

At any rate, every word can be clearly made out, though Foreman's characters do tend, like Foreman, to withhold their data from examination. Much of the libretto pays compliment to Thomson's vis-à-vis Gertrude Stein, with the gerund phrases ("Asking them to change their mind to accommodate changing places") and those helpful explanations ("We are not singing, we are only rolling our eyes") that resist exegesis but remain in the memory. Why all the mystery? Yet, on the other hand, why not? *Elephant Steps* is one of those "theatre pieces," mayhap, but its musical thrust is much firmer than those limp lays of love and death that we left straggling in the past like candies bitten in the box. A novelty *Elephant Steps* certainly is, all the more so sharing a chapter with *The Rake's Progress* and *Die Soldaten,* a "staged action," but not one that protests too much. It seems to know that it is not in "form" for "opera," but knows as well that form can make room for form, not forgiving but elaborating. What is permitted teaches us nothing.

Is there dissolution in this communicative but not sufficiently impartive piece, intransigent in its Steinese verse and a croaking chorus in its remorseless pastiche? Is it *Zeitoper* again—giggling at the gullibility of fashion, flensing the associational process from the theatre ritual like the flesh from off the Bacchae's quarry? For we know this is the day of much de-formed chicanery, of *Votre Faust* and *Reconstructie . . .* silly stuff, not worth befouling the chronicle with but for the pleasure of goading fashion's friends, the critics. Is that it, then?

No. Like another suspicious situation of the day, *Recital I (For Cathy)*, Luciano Berio's grotesque arrangement of a song recital for the singer Cathy Berberian—a *Liederabend* dedicated to Godot, complete with aleatory patter and personal confession—*Elephant Steps* proves that the chasm separating romantic expansion and neoclassical diligence has been bottled up, at least on certain stages. What at first glance appears to be an encounter with the operatic equivalent of untranslated Choctaw is not really so enigmatic. In the city of operas, Silverman and Foreman raised up a dwelling of some reference to other houses on the block, and moving down the thoroughfare to the older addresses, one finds references in their likenesses, too, pediment and obelisk. However mysterious it seems, *Elephant Steps* has its roots in experiments already presented herein, such as Malipiero's "comedies" (another descendant of Malipiero survives less mysteriously in the heart-on-sleeve *satura* of Argento's *Postcard from Morocco*). Oh, these puppets, the gay primitives, screaming because their heads are in the noose! Don't they know it's all in fun, and all for extrapolation? Don't they know they *teach?* The dramatic, for Malipiero, was what one sees, and the music had to express what one did not see; for Foreman, the dramatic simply joins the music in putting the unseeable on display.

Berio, too, pokes behind the mask in *Recital I,* with Berberian's help, letting her both *play* a singer and *be* Cathy Berberian, the effect of which on stage is akin to watching Cathy Berberian pretend she isn't pretending she isn't Cathy Berberian—yes, exactly! It doubles back on itself, thus acclimatizing the autotelic fantasy of theatre to the discomforts of the life-bag, the "real life." Likewise, *Elephant Steps*, though it admits of a cast of actors, roots into the fantasy valuation in fluent pastiche-speak. When the Ragtime Lady inveigles Hartman with "Watch me put my right foot through the door," we hear pop music, sophisticated so as to perplex us with surmise—just as they were doing in the days of Satie. Now, however, the quotation is more than intrusion; it amortizes the evening, and there is nothing in *Elephant Steps* but quotation, jam without the bread, so to speak. Having got *Four Saints in Three Acts*, *Il Prigioniero*, and *The Knot Garden* under our belts, however, we can make a place for *Elephant Steps* there, too, for the nonlinear narrative is now in its mannerist phase and the gyres are turning again.

With that informality that we spot in other American works, *Elephant Steps* is neoclassical form distended, informality being the tintype of the profane—and the sacred, for once, nowhere in sight. Henze played both ends across the middle in *Der König Hirsch* so as to retrieve his myth, but Silverman and Foreman's myth is informal, a beggar's myth. In their *Hotel for Criminals* (1974), an adaptation of Louis Feuillade's silent-movie serials of the Paris underworld, the commedia of doorways and props was more pronounced than in *Elephant Steps*—but if Punch and Judy are to lead the dance, whither form? Are these truly operas, or mere sybaritic sophistry disguised as art? (Are they still asking those tedious questions?) Still, what's new about mannerism? The Camerata got together on Arianna and Eurydice at the close of a mannerist era, in fact—so much for the theory that mannerism degrades form. And can one evade the lurking suspicion that atonality is the most flagrant mannerist gospel of all—that Schoenberg's suggestion that triadic harmony is alphabet soup and no single note matters more than another is not reinvention but dissolution?

None of this matters; *Elephant Steps* works. And here is what it is: the dramatic equivalent of atonalism, a horizontal rather than a vertical voyage of moments and a genre that owes its rhythm to expressionism, which in turn pays its due to comedy, which, we know, is forever changeless. In and around the lineage of the Word, music thrived, but now the two are in alignment: sacred, profane ———→ Godot. *Elephant Steps* belongs to Godot, for though it is music and words, they assault the senses in contraposition, each reminding the spectator of the other, expressing themselves rather than an entity; yet, as themselves, they add up to an entity together, tangibly different from and so clearly born of what used to be opera. Thematically, Godot inclines to Punch and Judy commedia, to the natural comity of the *gai primitif* and the incursions of *logos;* such is the age, and so be it. We've had warning enough from sources too various to distil as a movement, but the feeling has been in the air—the shrill, barbarian lotto of *Renard,* the coming to terms of Zerbinetta and Ariadne, the revival of the masks by Mascagni and the murder of them by Malipiero, the awful sorites of *Oedipus Rex*—this, therefore that, therefore annihilation—the cool plashing of Satie, the confident untowardliness of jazz opera, the baldly formalized augmentation of the tone row, the heartless doodle-doo of the golden cock, the Goethe quotations of the young lord, the flight of the nose, the guignol extravagance of the bassarids . . . all tutelary exercises branded with the poker of nihilism, for logic is nihilism to the romantic, and opera—let us make no mistake—is romantic at the core. It is myth, theatre myth, potent with the story. *Elephant Steps* waxes vague on the question of plot, yet it, like *The Rake's Progress* and *Der König Hirsch,* undertakes the quest. Stravinsky took his trip in rejuvenated format, sometimes sniggering, sometimes obviously moved by the eloquence of the plan; Henze's incantatory pilgrimage offers a prime illustration of how the ceremonial has taken the commedia under its wing—but where Henze's clown-alchemists are essentially as much on the outskirts of his show as the old ballad comedies were hunted off the main city stages to the fairgrounds, Silverman's people are all clowns; *Elephant Steps,* the puppet play updated, numbers grotesques of all kinds in its college of alchemy.

Godot has no heart, but some of his dependents reap the benefit of musical penetration even as they cling to the Word. Like the King-Stag they can both clarify themselves and stimulate with song. It's tricky to do, but then these are tricky times. We might have rung down the curtain on Henze's *Märchenoper mit Clowns* but for its derivation from old Gozzi; something tasting more of the neoteric, of modern rather than ancient absurdism, seems required. *Der König Hirsch* is fantasy. For the 1970s, we requisition the surreal. And so, for a finale: Gordon Crosse's *The Story of Vasco* (1974), based on Georges Schehadé's weird play about a meek little man accidentally turned into a hero.

Vasco, his story. Tom Rakewell, the King-Stag, Marie Wesener, Hartman . . . now it is the turn of a totally innocuous village barber, not the sort of chap you'd ever notice, who is enlisted by a general to run a secret mission during wartime. "What *is* a hero?" asks the general:

> A hero is a soldier who succeeds . . . is he not?
> But the shouting fiery fury,
> The fearless ferocious frightener,

He commits the fatal error—for he fails.
He attracts a little bullet—and he falls.
And lies forever dead among the other fools.
A hero is a soldier who succeeds . . . is he not?
The fearful creeping creature,
The butterfly, the frightened snail,
He's the one, he's the hero—every General must agree.
He gets through where all the others drop and cry.
And carries out his orders while the others feed the crow.

Crosse's polytonal idiom, occasionally resorting to serialism, points up the absurdity of heroics by letting the general sing the "butterfly" lines to inversions of the "fury" lines, for only the reverse-hero, the *Hanswurst,* can carry the flag in seriocomedy. Oh, all the earmarks of the absurd theatre are here—the comic interludes and oddballs; a chorus of anthropomorphic crows keen for carnage; non sequiturs; garish goings-on in all seriousness; and a romance between two people who only meet once, and who love only ideals of each other. "What *is* a hero?"—a very appropriate question. What is a hero nowadays? The rake-hell Rakewell, the bewildered Prisoner, the organic twins Pentheus and Dionysus, all gave us changeling heroes, yet the moral imperatives that Western theatre grew up on obtain even so. The heroic expedition has not changed; the hero has. And so has the expeditor.

Musically, Crosse gives us two Vascos. One is the childlike innocent, characterized by a perky diatonic march, the other a nobler fellow, sounded in a slow theme calmly jumping wide intervals, hero music. The first opens the second act (to accompany a drop curtain representing Vasco's offensive as one of those maps such as confirm the eidolon of Tolkien or Oz books). The second is heard when Vasco meets up with the girl who loves him and whose image in turn fires his courage; no more than a

moment in a frantic burlesque, it counts for everything even
though the girl does not realize who he is until he has gone, for
it is not dreams that supplement life, but the reverse.

It might as well be so, if the dream is so lively as here. That
chorus of crows, for example, is costumed not unlike a cell of
monks, and they open the opera, aided by the wind machine and
a bit of static percussion, in a wordless, nasal chant directly
aimed at medieval plainsong. Not till they have finished does
the orchestra enter, and this is a very atmospheric beginning,
draining the Gregorian image of its security and contemplation.
One expects a first-class theatrical effort from the modern composer, what with all the effortful materials at his disposal, but
Crosse is a melodist of rank, and no sooner has Lieutenant September marched in to curse the villain crows than a flowing $\frac{5}{4}$
movement exhales cantilena into the proceedings. "And tonight
is like autumn, when the world tries to scour itself clean with its
rainy gales."

The libretto of *The Story of Vasco*, adapted by the composer
from Ted Hughes' loose translation of Schehadé, is one of the
most resourceful *versions* of an already dramatic situation, one of
those operas-from-plays that make revivals of the play a flat and
dispirited affair ever after. One always wants to know how acts
begin and end, what pins the composer selects for the arresting
of his pictures—this being one of the files in the drawer marked
Theatre, and having much to do with the conjunction of sacred
spell-casting and the profane klaxon. For example, toward the
close of the second act, the girl, Marguerite, who loves the Vasco
she has never seen, simple barber that he is, reflects on the wisdom of her passion in long, lyrically declarative lines, each one
echoed by three solo violins in turn. "But why do those lovers

who hold each other keep their eyes closed?" she asks. "What else do they hold but the dream of their love?" And then, momentously, Vasco appears, and Marguerite sees him at last, at last! But there is nothing momentous about it, no at last, and the orchestra moves on to other matters without comment, for Marguerite and Vasco, like Pyramus and Thisby, are all eyes open and seeing: they haven't taken hold of the dream yet.

This is indeed subtle scoring. Instead of grimacing So We'll Know, Crosse goes right into a duet for these two strangers, starting with a charming solo for Marguerite, "The first time I saw him," describing her "Vasco" to Vasco. Crosse can sound as discombobulated as any revolutionary when he likes to, but here a most highly accessible communication delivers the feint of aria, with the development and recapitulation of figures that we have learned to learn from in opera. Marguerite invents a meeting with her hero: "He waved, and the horse waved. I shut the window so I should see nothing else all the rest of the day." As Vasco and Marguerite converse, her obsession begins to impress her and now, only now, the music is willing to climax—for this is the dream, you see?—and Marguerite builds up on repeated high B's to the summit of the scene: "I love him too much, I love him too much, I cannot bear to think of him." And then, quietly: "He went away . . . I've never seen him since." And then: "To tell you the truth—I've never seen him." And then most of all, the two look at each other. Still not grimacing, but with a reasonable momentousness, the orchestra tells us what we want to hear— that she is almost seeing him—and, not having connected fantasy to truth, the two lovers part as Vasco goes off on his mission, unrecognized. But Marguerite notices that the fellow has dropped a pair of scissors. "Vasco! It was Vasco!" she cries, all on one note. Others restrain her from dashing after him, and a snare drum cuts into the excitement. The characters freeze; one of them comes downstage and announces, "The position is as follows: Black!" And the act is over, capped by a loud button in— well, of all things!—C Major.

Eventually Vasco does become a hero. Inspired by his love for Marguerite, he loses the innocence that has protected him, pulls the coup that wins the war for the general . . . and loses his life. The crows would seem to have won, but, as the curtain falls, we hear Vasco's two themes played together, the diatonic march and the full-bodied hero's theme—for, ultimately, Vasco was a hero: he became what Marguerite dreamed.

Altogether wise, witty, poetic, and dramatic, *The Story of Vasco* is a fine figure of modern opera, eminently suitable to officiate at a closing paragraph. Bimodal in the age of Godot, opera has combined the spiritual benison of the sacred with the cutting edge of the profane. Naturalism now mates with romance, heroism adapts to comic logic, parody controls the spectator's flight of belief, and expression may disagree, contrapositively, with action. Most importantly, the Word stands in relief, for where the music drama of Gluck and Wagner sought a primarily musical context, modern opera accentuates the script. And yet, admitting all this, one must admit as well that the music of it has the same power to move that music has always had even as it tears the veil off of text. Music may be for the inexpressible, but opera has a story to tell, and "story" is, as has been said, largely a matter of instinct. Opera, too, then, is instinctual—song with a kinetic face, theatre with ulterior dimensions. We must hear the words as well as the music, but we *shall* have the music; and there will be ideas as well, or what *is* a hero for?—and what for, his myth? Our long itinerary through to the art of Godot should have answered such questions—doubters may apply to Godot himself for reassurance. Wait around a bit; he won't be long.

Index

Abandon d'Ariane, L', 213
Abraham, Gerald, 194
Adami, Giuseppe, 88-89, 223
Adriana Lecouvreur, 258
Aeschylus, 7, 152
Ägyptische Helena, Die, 235, 239, 240
A Kékszakállú Herceg Vára, 38-39, 110, 204, 206, 301, 326
Albéniz, Issac, 68
Albert Herring, 208, 276, 289, 290-91, 294
Alfano, Franco, 88, 94, 105, 173, 222-23, 229
Amadis, 54
Amahl and the Night Visitors, 231
Amelia al Ballo, 230, 232
Amica, 78
Amico Fritz, L', 73
Amor Brujo, El, 68
Amore dei Tre Re, L', 95-97, 99
An Allem Ist Hütchen Schuld, 107
Anderson, Maxwell, 320
Andrea del Sarto, 220
Angelique, 215
Aniara, 205
Animal Crackers, 6
Annonce Faite à Marie, L', 229
Antar, 62, 129
Antheil, George, 284, 302
Antigonae, 253-54
Antigone, 65, 105, 211
Antonio e Cleopatra, 226
Antony and Cleopatra, 310-11
Apollinaire, Guillaume, 216-18
Apollo and Marsyas, 207
Arabella, 115-16, 235-37, 266
Argento, Dominick, 317, 339

Ariadne auf Naxos, 40-45, 112, 119, 120, 122, 124, 237, 239, 241
Ariane (Martinů), 190-91
Ariane (Massenet), 58
Ariane et Barbe-Bleue, 29, 30, 60, 62, 131, 133
Arlecchino, 130-31
Arme Heinrich, Der, 111
Assassinio nella Cathedrale, L', 224-25
Athaliah, 309
Atlántida, L', 206
At the Boar's Head, 275
Auber, Daniel-François-Esprit, 12, 263
Auden, W. H., 172, 265-66, 267-69, 287, 326-29
Aufstieg und Fall der Stadt Mahagonny, 144-48
Aventures du Roi Pausole, Les, 65-66

Bacchus, 58
Bachmann, Ingeborg, 264, 266
Bálazs, Béla, 38-39
Ballad of Baby Doe, The, 312-14
Ballet Ballads, 180-81
Barber, Samuel, 309-11
Barbier, Jules, 30
Bardi, Giovanni de', 3
Bärenhäuter, Der, 107
Barkworth, John Edmund, 273
Bartók, Béla, 38-39, 110, 191, 204, 206, 301
Bassarids, The, 9, 168, 191, 239-40, 267-70, 271, 311, 325
Beatrix Cenci, 205-6
Bécaud, Gilbert, 52, 220-21, 256
Beeson, Jack, 309

347

Belasco, David, 79, 84
Bellini, Vincenzo, 11
Bellow, Saul, 315
Bell Tower, The, 246
Benelli, Sem, 77, 95-96, 97
Bennett, Richard Rodney, 284-86
Berberian, Cathy, 339-40
Berg, Alban, 77, 93, 126, 131, 137, 138, 158-62, 173, 226, 243, 247-52, 334, 335
Berio, Luciano, 232, 339-40
Berkeley, Lennox, 277
Berlioz, Hector, 10, 13, 36
Bernanos, Georges, 218
Bernstein, Leonard, 299, 319-20
Besuch der Alten Dame, Der, 254-56
Betrothal in a Monastery. See *Obrucheniye v Monastirye*
Bilby's Doll, 311
Billy Budd (Britten), 170, 291-92
Billy Budd (Ghedini), 292
Birtwhistle, Harrison, 282-83
Bizet, Georges, 317
Blacher, Boris, 255, 261, 262
Bliss, Arthur, 277
Blitzstein, Marc, 299, 318, 319
Bloch, André, 174
Bloch, Ernest, 62
Blomdahl, Karl-Birger, 204-5
"Blonda, Max" (Gertrud Schoenberg), 139
Blue Monday, 148, 302
Bluthochzeit, Die, 207, 259
Boito, Arrigo, 105-6, 171
Bolivar, 214
Bomarzo, 205-6
Bond, Edward, 271
Borgese, G. A., 315
Boulevard Solitude, 263, 264
Brebis Egarée, La, 212
Brecht, Bertolt, 126, 144-48, 196, 215, 245, 259-60
Britten, Benjamin, 78, 81, 93, 170, 171, 191, 252, 262, 276, 278, 286-98, 299, 311, 326
Brontë, Emily, 311
Büchner, Georg, 36, 137, 158, 160-61, 204, 254, 334
Bungert, August, 50, 107-8

Burning Fiery Furnace, The, 294
Busenello, Giovanni Francesco, 9, 208, 325
Busoni, Ferruccio, 14, 47-48, 77, 88-90, 91, 130-33, 147, 151, 235, 247, 255, 294, 330
Byelsky, Vladimir, 32, 33

Cable, George Washington, 274
Cadman, Charles Wakefield, 301
Cagliostro, 224
Cain, Henri, 56
Campiello, Il, 91
Candide, 319-20
Čapek, Karel, 173, 184, 187
Cappello di Paglia di Firenze, Il, 230
Capriccio, 125, 240-43, 256, 326
Captain Jinks of the Horse Marines, 309
Cardillac, 105, 129-30, 131, 139, 149, 244, 267, 311
Carée, Michel, 30
Carmen, 53, 317
Carmen Jones, 317
Carner, Mosco, 150
Cavalieri di Ekebù, I, 100
Cavalleria Rusticana, 20, 73, 81
Celestina, La, 68
Cena delle Beffe, La, 77
Cendrillon, 21, 53
Chadwick, George Whitefield, 299
Charpentier, Gustave, 19-24
Chausson, Ernest, 52
Chérubin, 56-58
Chien du Régiment, Le, 64
Christelflein, Das, 111
Christophe Colomb, 212, 214-15, 223, 252
Claudel, Paul, 211-12, 213, 214
Cléopâtre, 54-55
Cocteau, Jean, 64, 65, 151, 213, 219
Colas Breugnon. See *Mastyer iz Klamsi*
Columbus, 252
Comedy on the Bridge, 190
Conchita, 100
Conrad, Joseph, 285
Consul, The, 230-32

Converse, Frederick Shepherd, 300
Copland, Aaron, 307
Cordovano, Il, 229
Crabbe, George, 287
Cradle Will Rock, The, 319
Cramer, Heinz von, 255, 264, 329-33
Crepusculum, 77
Cristoforo Colombo, 72
Cross, Beverley, 282, 285
Crosse, Gordon, 342-46
Crozier, Eric, 290-92
Crucible, The, 309
Cunning Little Vixen, The. See
  Příhody Lišky Bystroušky
Curlew River, 294
Cyrano de Bergerac (Alfano), 173,
  222-23, 229
Cyrano de Bergerac (Damrosch), 303

D'Albert, Eugen, 108-9
Dallapiccola, Luigi, 223, 226-28, 229,
  235, 257
Dama a Lupici, 208
Damase, Jean-Michel, 220
Damrosch, Walter, 302-3
D'Annunzio, Gabriele, 28, 61, 74-75,
  82, 93, 97-98, 99, 103, 104, 149,
  174, 222
Dantons Tod, 254-56, 262
Daphne, 138, 239, 240
Da Ponte, Lorenzo, 9, 115
Das Kommt Davon, 247
David, 214, 215
Davies, Peter Maxwell, 296
Death in Venice, 191, 289, 293-94,
  296, 297-98
Débora e Jaèle, 103-4
Debussy, Claude, 24-30, 36, 50, 61-
  62, 64, 65, 75, 149, 159, 162, 174,
  181, 190, 204, 211, 216, 224, 235,
  252, 258, 278
Decembrists, The. See *Dyekabristi*
Delannoy, Marcel, 215
Delius, Frederick, 30, 273-75, 276
Déliverance de Thesée, La, 213
Dessau, Paul, 259-60
De Temporum Fine Comoedia, 254
Devil and Daniel Webster, The, 305-6
Dialogues des Carmelites, 218-19

Diary of a Madman, 285
Dibuc, Il, 223
Doktor Faust, 130, 131-33, 251, 282,
  294
Don Giovanni, 9, 58, 127
Don Juans Letztes Abenteuer, 109
Donizetti, Gaetano, 11, 259
Donna Uccisa con Dolcezza, Una, 208
Donne Curiose, Le, 91
Don Quichotte, 58-59, 69, 220
Don Rodrigo, 205
Dottor Antonio, Il, 222
Down in the Valley, 321
Dreigroschenoper, Die, 105, 144-48,
  259, 260, 319, 320
Drömmen om Thérèse, 205
Dukas, Paul, 50, 60, 75, 131, 216, 227
Duke Bluebeard's Castle. See *A
  Kékszakállú Herceg Vára*
Dupont, Gabriel, 62
Dvořák, Antonín, 186-87, 303
Dyekabristi, 202
Dzerzhinsky, Ivan, 200

Eccentricities of Davy Crockett, The,
  180-81
Edipo Re, 95
Egk, Werner, 252-53
Einem, Gottfried Von, 192, 254-57,
  262, 295
Einstein, 260
Elegy for Young Lovers, 265-66, 270
Elektra, 40, 80, 115-17, 128, 239, 240
Elephant Steps, 337-42
Eliot, T. S., 222, 224
Elisabeth Tudor, 259
Emperor Jones, The, 144-45, 302
Ende Einer Welt, Das, 263
Enfant et les Sortilèges, L', 144, 216
Enlèvement d'Europe, L', 144, 213
Eroi di Bonaventura, Gli, 234
Ersten Menschen, Die, 109-10
Erwartung, 127, 155-57, 161-62, 169,
  247
Escorial, 309
Étranger, L', 52
Euménides, Les, 152
Euripides, 8, 267-70
Eurydice (Damase), 220

Façade, 140-41, 151, 284
Falla, Manuel de, 68-70, 173, 206
Falstaff, 9, 11, 15, 36, 71, 87, 91, 118
Fanciulla del West, La, 84-86, 87, 103
Farberman, Harold, 315
Fauré, Gabriel, 27, 52, 60-61
Favola del Figlio Cambiato, La, 226
Fedra, 103
Fennimore and Gerda, 275, 279
Fervaal, 52
Feuersnot, 45, 112-14, 165
Feuillade, Louis, 340
Février, Henri, 60
Fidelio, 130, 239
Figlia di Iorio, La, 224
Flaming Angel, The. See *Ognyeniy Angyel*
Flower and Hawk, 311
Floyd, Carlisle, 311-12, 313
Foreman, Richard, 338-40
Forster, E. M., 292
Fortner, Wolfgang, 207, 259
Forzano, Giovacchino, 87, 92, 95
Foss, Lukas, 308
Four Saints in Three Acts, 303-6, 307, 340
Fra Gherardo, 103-4, 105
Francesca da Rimini, 54, 99-102, 304
Franchetti, Alberto, 70, 71-72
Franc-Nohain, 66-67
Françoise de Rimini, 101
Frau Ohne Schatten, Die, 120-23, 124, 279
Frédégonde, 52
Fried, Barbara, 315
Friedenstag, 239, 240
From the House of the Dead. See *Z Mrtvého Domů*

Gambler, The. See *Igrok*
Ganne, Louis, 67
Gargiulo, Terenzio, 229
Gatty, Nicholas Comyn, 273
Germania, 72
Gershwin, George, 148, 176-77, 181, 302, 303-5
Ghedini, Giorgio, 292
Ghelderode, Michel de, 24, 309

Ghizlanzoni, Alberto, 173
Giacosa, Giuseppe, 80
Gianni Schicchi, 87, 88, 92, 174
Gide, André, 41, 111, 149, 220
Gilbert, Anthony, 283
Ginastera, Alberto, 205-6
Ginevra, 215
Gioielli della Madonna, I, 90
Giordano, Umberto, 70, 76-78
Giovanni Gallurese, 95
Giovanni Sebastiano, 233
Gismonda, 60
Giulio Cesare, 226
Giulietta e Romeo, 78, 102
Gloriana, 289, 293, 300
Gluck, Christoph Willibald von, 8, 9, 11, 52, 59, 63, 139, 155, 161, 165, 208, 219, 221, 241, 346
Glückliche Hand, Die, 155-57
Gogol, Nicolai, 285
Golden Apple, The, 319
Golden Cock, The. See *Zolotoy Pyetushok*
Goldoni, Carlo, 46, 90, 91, 92, 151, 153
Golisciani, Enrico, 170
Goyescas, 68
Gozzi, Carlo, 46-48, 88-89, 130, 173, 264, 329-33, 337
Graener, Paul, 109
Granados, Enrique, 68
Green, Paul, 245
Gregor, Josef, 239-40
Gruenberg, Louis, 144-45, 302
Guignol, 174
Guiraud, Ernest, 52
Guntram, 113
Gurrelieder, 36-37, 114-15, 141-42, 154, 186

Haas, Joseph, 261
Hába, Alois, 189
Halffter, Ernesto, 206
Hamilton, Iain, 337
Hamlet (Searle), 207, 285
Hamlet (Szokolay), 207
Hamlet (Thomas), 30
Hänsel und Gretel, 107
Happy End, 145

Harmonie der Welt, Die, 244-45
Harte, Bret, 189
Háry János, 207
Hassall, Christopher, 277
Hazon, Roberto, 208
Hélène, 59, 60
Hellman, Lillian, 318, 320
Help! Help! The Globolinks!, 232
Henneberg, Claus, 258
Henze, Hans Werner, 82, 183, 191, 192, 252, 262-73, 288-89, 336
Herbert, Victor, 299, 301, 308
Hero, The, 232
Herr Puntila und Sein Knecht Matti, 259
Herr von Hancken, 205
Herz, Das, 111
Heure Espagnole, L', 66-67, 216, 284, 304
He Who Gets Slapped, 309
Heyward, Dorothy, 177
Heyward, DuBose, 177
Hindemith, Paul, 127, 128-30, 131, 139-40, 144, 151, 244-45, 247, 267
Hin und Zurück, 127, 130, 144
Histoire du Soldat, L', 23, 141-42, 149, 162
Hochzeit des Jobs, Die, 261
Hofmannsthal, Hugo von, 40-45, 115-24, 235-37, 240, 278, 282
Hoiby, Lee, 309
Holst, Gustav, 275, 276
Homerische Welt, 50, 107-8
Honegger, Arthur, 65-66, 211-12
Hotel for Criminals, 340
Hughes, Ted, 344
Hugh the Drover, 276
Humperdinck, Engelbert, 107, 162, 303
Hurník, Ilja, 208

Ibert, Jacques, 215, 216
Ibsen, Henrik, 252
Ifigenia, 224
Igrok, 173, 194-95, 208
Ilias, Das, 107
Illica, Luigi, 72, 73, 74, 80, 153, 300
Incoronazione di Poppea, L', 9, 238, 325

D'Indy, Vincent, 21, 52, 110
Ines de Castro, 229
In Seinem Garten Liebt Don Perimplin, 259
Intermezzo, 124-25, 139, 235, 237
Intolleranza 1960, 232
Iphigénie en Aulide, 241
Iphigénie en Tauride (Gluck), 8, 138-39
Iphigénie en Tauride (Piccinni), 8
Iris, 72, 73
Irische Legende, 252-53
Isabeau, 74

Jacobs, Arthur, 201
James, Henry, 111, 294, 296, 333
Janáček, Leoš, 75, 76, 129, 141-42, 162, 173, 177, 181-89, 204, 329-33, 340, 342
Jarnach, Philipp, 131
Jasager, Der, 245
Jeanne D'Arc au Bûcher, 211-12
Její Pastorkyna, 182-84, 186, 204, 220, 266
Jenůfa. See *Její Pastorkyňa*
Johnny Johnson, 245
Jongleur de Notre-Dame, Le, 55-56, 220, 258
Jonny Spielt Auf, 83, 105, 128, 142-44, 181, 215, 247, 302, 325
Jonson, Ben, 237-38, 276
Joplin, Scott, 175-76
Judith, 65
Julien, 23-24
Julietta, 189-90
Jumping Frog of Calaveras County, The, 308
Junge Lord, Der, 266-67, 270, 291

Kabalyefsky, Dmitri, 202-3
Kallman, Chester, 172, 265-66, 267-69, 326-29
Karl V, 138, 245-46
Kashchey Byesmyertni, 31, 33
Kâta Kabanová, 129, 185-86
Katyerina Ismailovna, 200-201
Kern, Jerome, 176
King Priam, 168, 258, 279-81, 290, 333

King's Henchman, The, 303
Kirchner, Leon, 315-16
Kirke (Bungert), 107
Kirke (Egk), 252, 253
Kluge, Die, 178-79
Knot Garden, The, 280, 281, 340
Knudsen, Poul, 127
Koanga, 274-75, 276
Kodály, Zoltan, 207
König Hirsch, Der, 264, 265, 267,
    311, 329-33, 337, 340, 342
König Nicolo, 258
Königskinder, 107
Korngold, Erich Wolfgang, 109-10,
    243
Krauss, Clemens, 240-41
Křenek, Ernst, 14, 126, 127-28, 137,
    138, 142-44, 150, 162, 181, 243,
    245-47, 294
Król Roger, 191, 207

Lachmann, Hedwig, 114
Lady in the Dark, 317
Landré, Guillaume, 220
Lanzelot, 260
Larson, Jack, 316
Last Savage, The, 231-32
Latouche, John, 180-81, 312-14, 319,
    320
Leben des Orest, 128, 223
*Legend of the Invisible City of
    Kityezh, The.* See *Skazaniye . . .*
Leggenda di Sakùntala, La, 88
Lenz, Jakob Michael Reinhold, 160-
    61, 334-35
Leoncavallo, Ruggiero, 70, 76-77, 95
Lesur, Daniel, 220
Levadé, Charles, 215
Levy, Martin David, 309
Lidé z Pokerflatu, 189
Liebe der Danae, Die, 239-40
Liebermann, Rolf, 258, 261-62
Lily, 315-16
Lion, Ferdinand, 130
Little Sweep, The, 291
Lord Byron, 316
Losers, The, 315
Lost in the Stars, 320-21
Louise, 9, 19-24, 81, 125, 257, 337

Love for Three Oranges, The. See
    Lyubof k Tryem Apyelsinam
Love's Labour's Lost, 172, 174
Lully, Jean-Baptiste, 7, 165, 194, 208,
    216, 219
Lulu, 138, 162, 247-52, 257, 328, 329
Lyedy Macbyet Mtsenskowo Uyezda.
    See *Katyerina Ismailovna*
Lyubof k Tryem Apyelsinam, 194-97,
    198

Macbeth (Bloch), 62
Madama Butterfly, 78, 79-84, 85
Madetoja, Leevi, 173
Madonna Imperia, 105
Maeterlinck, Maurice, 24-30, 60, 233,
    282
Makropulos Business, The. See *Věc
    Makropulos*
Mala Vita, 76
Malheurs d'Orphée, Les, 212-13
Malipiero, Gian Francesco, 94, 150-
    51, 153, 222, 225-26, 229, 233-34,
    294, 339, 341
Mamelles de Tiresias, Les, 216-18
Mann, Thomas, 13, 278, 297
Manon Lescaut (Auber), 12
Manon Lescaut (Puccini), 78
Man Without a Country, The, 303
Maria Golovin, 231
Marie Antoinette, 229
Marina Pineda, 220
Mariotte, Antoine, 62
Maroûf, Savetier de Caire, 54, 62
Martinnen, Tauno, 207
Martinů, Bohuslav, 189-91, 194
Martyre de Saint-Sébastien, Le, 28,
    36-37, 61-62, 67, 211
Mascagni, Pietro, 70, 73-76, 78, 80,
    86, 95, 103, 153, 174, 318, 341
Maschere, Le, 72, 73-74
Maskarade, 153, 204, 207
Massenet, Jules, 21, 30, 50, 53-59,
    69, 70, 120, 183, 212, 219-20, 258,
    263, 337
Mastyer iz Klamsi, 202-3
Mathis der Maler, 138, 165, 244-45
Matka, 189
Maw, Nicholas, 282

Maximilien, 214
Medium, The, 230-31
Meistersinger von Nürnberg, Die, 6, 9, 36, 80, 91, 118, 244
Melusine, 258-59
Melville, Herman, 246, 289, 292
Mendès, Catulle, 58
Menotti, Gian Carlo, 230-32, 252, 309-10, 317
Mère Coupable, La, 212, 215
Messager, André, 67
Metamorfosi di Bonaventura, Le, 234
Midsummer Marriage, The, 279-80, 281
Midsummer Night's Dream, A, 289, 294-95
Milhaud, Darius, 82, 137, 138, 144, 152-53, 173, 212-15, 294
Millay, Edna St. Vincent, 303
Miller, Arthur, 229, 309
Molière, 40-41, 262
Mona, 300
Mond, Der, 178-80
Moniuszko, Stanislav, 191
Monna Vanna, 60
Montemezzi, Italo, 94, 95-99, 174, 222
Montezuma, 315
Moore, Douglas, 252, 305-6, 312-14, 338
Moross, Jerome, 180-81, 319
Morte dell'Aria, La, 229
Morte delle Maschere, La, 153
Moses und Aron, 162, 247-52
Moskva, Cheryomushki, 200
Most Important Man, The, 232
Mother of Us All, The, 306-8
Mozart, Wolfgang Amadeus, 9, 10, 14, 43, 91, 115, 118-19, 130, 131, 215, 238, 262, 266, 300, 308, 325, 326-27, 333
Mozart i Salieri, 31
Much Ado About Nothing, 273

Nabokov, Nicholas, 172
Natalia Petrovna, 309
Natoma, 301
Nausikaa, 107-8
Nave, La, 97-99, 174, 177

Negri, Gino, 233
Nelson, 277
Nerone, 78, 105-6
Nestroy, Johann, 262
Neues vom Tage, 139, 156
Nielsen, Carl, 153, 204
Nino, 63, 216
Nono, Luigi, 232
Nos, 201-2
Nose, The. See Nos
Noye's Fludde, 294
Nozze di Figaro, Le, 118, 121
Nusch-Nuschi, Das, 127

Obrucheniye v Monastirye, 198
Odysee, Die, 107-8
Odysseus Heimkehr, 107-8
Odysseus Tod, 107-8
Oedipus der Tyran, 253-54
Oedipus Rex, 151-52, 204, 328
Offenbach, Jacques, 14, 67, 143
Of Mice and Men, 311, 313
Ognyeniy Angyel, 198
Oliver, Stephen, 170-71
Olympians, The, 277
O'Neill, Eugene, 144, 309
Opera, 232
Opéra d'Aran, 52, 220, 256, 258, 318
Oresteia, The, 7
Orestie, L', 137, 152-53, 214
Orfeide, L', 150, 162, 233
Orfeo, ovvero L'Ottava Canzone, 153
Orff, Carl, 152-53, 178-180, 243, 253-54
Oro, L', 224
Orséolo, 224
Ostrofsky, Alyexandr, 185
Our Man in Havana, 283, 286
Owen Wingrave, 296-97

Pacific Overtures, 176, 299
Padmâvatî, 62, 129
Pagliacci, 20, 75, 76
Palestrina, 111-12, 132, 166, 265
Pallas Athene Weint, 239, 246
Panurge, 54
Parisina, 73, 74
Parker, Dorothy, 320

Parker, Horatio, 300
Parsifal, 52, 80, 249
Pasatieri, Thomas, 229, 313
Passaggio, 232, 271
Paul Bunyan, 287
Pauvre Matelot, Le, 213-14
Pears, Peter, 294
Pedrell, Felipe, 68
Peer Gynt, 252
Pelléas et Mélisande, 13, 23-29, 30, 62, 65, 67, 95, 204, 208, 219, 229, 257, 258, 309, 325
Penderecki, Kryzstof, 191-92, 295
Pénélope (Fauré), 60-61
Penelope (Liebermann), 239, 258
Penny for a Song, A, 284
Perfect Fool, The, 275
Pergolesi, Giovanni Battista, 7, 67, 91, 130, 194, 208
Persée et Andromède, 216
Perséphone, 36-37, 149
Peter Grimes, 80, 220, 276, 278, 287-88, 289, 290, 292, 293, 337
Peter Ibbetson, 303
Petrassi, Goffredo, 223, 229
Pfitzner, Hans, 111-12, 130, 132, 166, 247, 255
Piccini, Niccolò, 8, 9, 46, 59, 238
Piccolo Marat, Il, 95
Pilgrim's Progress, The, 276
Pipe of Desire, The, 300
Piper, Myfanwy, 294, 296-97
Pirandello, Luigi, 226, 309
Pireneus, Els, 68
Pizzetti, Ildebrando, 94, 103-4, 110, 129, 174, 222, 223-25, 252
Plato, 65
Plomer, William, 293
Pohjalaisa, 173
Poltettu Oranssi, 207
Porgy and Bess, 176-77, 299, 303-6
Postcard from Morocco, 317, 339
Poulenc, Francis, 64, 216-19
Pousseur, Henri, 221-22
Povyest o Nastoyashchem Chelovyekye, 198-99
Preussiches Märchen, 261
Priestley, J. B., 277
Prigioniero, Il, 226-28, 231, 340

Příhody Lišky Bystroušky, 186-87, 304
Princesse de Chine, La, 47
Prinzessin auf der Erbse, Die, 144
Prinz von Homburg, Der, 264-65
Prodigal Son, The, 294
Prokofyef, Syergyey, 173, 192, 194-99
Prométhée, 60
Prometheus, 253-54
Prozess, Der, 254-55, 257
Pruslin, Stephen, 282
Pryeys, A. G., 200, 201
Puccini, Giacomo, 14, 47, 70, 78, 79-90, 92, 96, 103, 105, 110, 150, 173, 174, 183, 223, 233, 263, 300, 318, 330
Punch and Judy, 282-83, 337

Quattro Rusteghi, I, 90-91

Rabaud, Henri, 54, 62
Rake's Progress, The, 99, 148, 149, 168, 239, 262, 326-29, 333, 337, 338, 342
Ramuz, C. F., 142, 154
Rape of Lucretia, The, 78, 276, 289, 290, 291-92, 294, 299
Ratsumies, 207
Rautavaara, Einojuhani, 207
Ravel, Maurice, 52, 66-67, 75, 144, 216
Re Cervò, Il. See Der König Hirsch
Recital I (For Cathy), 339-40
Reconstructie, 272, 339
Red Rivals, The, 301
Regina, 318-19
Reimann, Aribert, 258-59
Reine Morte, La, 229
Renard, 154, 204, 328
Respighi, Ottorino, 94, 173
Retablo de Maese Pedro, El, 68-70, 206
Reutter, Hermann, 254
Revisor, Der, 252-53
Reyer, Ernst, 21, 50
Reznìček, Emil Nikolaus von, 110, 127
Riders to the Sea, 276

Rimsky-Korsakof, Nicolai, 31-35, 48,
53, 75, 97, 148, 166, 183, 193, 207
Ring des Nibelungen, Der, 32, 36,
54, 76, 80, 107, 113, 114, 159,
219, 333
Ring des Polycrates, Der, 109
Rising of the Moon, The, 282, 283
Risurrezione, 222
Ritter Blaubart, 110
Rivier, Jean, 215
Rocca, Ludovico, 223
Roi Arthus, Le, 52
Roi David, Le, 65
Roland von Berlin, Der, 77
Roma, 54
Romeo and Juliet, 273
Rondine, La, 86
Rosenkavalier, Der, 40, 45, 80, 110,
112, 115, 117-20, 121, 235-36,
241, 283, 304, 335
Rose vom Liebesgarten, Die, 111
Rossato, Arturo, 102
Rossellini, Renzo, 229, 256
Rossignol, Le. See Solovyey
Rota, Nino, 230
Roussel, Albert, 62-64, 110, 173, 215,
319
Royal Hunt of the Sun, The, 337
Rusalka, 186-87

Sadko, 31-32, 97, 166, 167
Sagredo, La, 223
Saint of Bleecker Street, The, 230-31
Saint-Saëns, Camille, 52, 59
Salammbô, 21
Sallinen, Aulis, 207
Salomé (Mariotte), 62
Salome (Strauss), 40, 80, 114-15, 116
Sancta Susanna, 127
Satie, Eric, 26, 52, 64-65, 69, 126,
173, 284, 340, 341
Sauguer, Louis, 220
Sauguet, Henri, 215
Saul og David, 204
Sāvitri, 275
Scene Machine, The, 283
Schehadé, Georges, 342, 344
Schoenberg, Arnold, 25, 27-28, 36-37,

114, 122, 123, 126, 127, 138, 139-
40, 154-57, 159, 161-62, 169, 186,
202, 221, 226, 243, 247-52, 263,
291, 335
Schule der Frauen, Die, 261-62
Schweigsame Frau, Die, 119, 236,
237-39
Seagull, The, 313
Second Hurricane, The, 306
Segreto di Susanna, Il, 91, 169-70
Serva Padrona, La, 7, 8, 67
Sessions, Roger, 260, 315
Sette Canzoni, 150, 225
Sguardo dal Ponte, Uno, 229, 256
Shakespeare, William, 30, 92, 171-72,
207, 273, 294-95, 330
Shanewis, 301
Shaporin, Yuri, 202
Sholokhof, Mikhail, 200
Shostakovitch, Dmitri, 192, 199-202
Show Boat, 176
Sì, 86
Sibelius, Jean, 27
Siberia, 72, 76
17 Tage und 4 Minuten, 253
Sigurd, 50
Silverman, Stanley, 337-42
Simoni, Renato, 88-89
Sir John in Love, 171-72, 174, 276,
279, 303
Sitwell, Edith, 140
Sitwell, Osbert, 140, 151
Six Characters in Search of an
Author, 309
Skazaniye o Nyevidimom Gradye
Kityezhe i Dyevye Fyevronii, 31-
33, 166, 167, 193, 207, 251
Skazka o Tsarye Saltanye, 193
Slater, Montagu, 193
Sly, 91-93, 105, 106, 233
Soldaten, Die, 160-61, 168, 333-37,
338
Solovyey, 54, 193-94, 204, 328
Sondheim, Stephen, 176
Spiel oder Ernst?, 127
Sprung Über den Schatten, Der, 127-
28, 137
Stanford, Charles Villiers, 273
Stein, Gertrude, 304, 306-8, 316, 338

Steinbeck, John, 311
Stephan, Rudi, 109-10
Stoba Komachi, 309
Story of Vasco, The, 342-46
Strauss, Richard, 40-45, 59, 87, 93,
    112-25, 126-27, 133, 139, 183, 233,
    235-43, 247, 278, 294, 326, 337
Stravinsky, Igor, 23, 25, 54, 123, 138,
    141-42, 148-49, 150, 151-52, 154,
    174, 191, 193-94, 204, 239, 252,
    262, 278, 326-29, 342
Street Scene, 319, 320
Strindberg, August, 309
Strobel, Heinrich, 262
Stronger, The, 309
Summer and Smoke, 309
Suor Angelica, 87, 174
Susanna and the Elders, 181
Susannah, 311-12, 313
Svanda Dudak, 189
Syemyon Kotko, 197
Symphonie Pastorale, La, 220
Syostri, 203
Székelyfonó, 207
Szokolay, Sandor, 207
Szymanowski, Karol, 191

Tabarro, Il, 70, 87-88
Tabasco, 299
Tamu-Tamu, 232
Tannhäuser, 50, 83
Taverner, 296-97
Taylor, Deems, 303, 308
Tempest, The, 273
Tender Land, The, 307
Tenor, The, 309
Testament de la Tante Caroline, Le,
    62-64, 215, 319
Teufel von Loudon, Die, 192
Thérèse, 30
Thomas, Ambroise, 20, 30, 53, 101
Thomson, Virgil, 303-5, 306-8, 316,
    338
*Threepenny Opera, The. See Die
    Dreigroschenoper*
Tiefland, 108-9
Tikhy Don, 200
Tippett, Michael, 278-81, 333
Toch, Ernst, 144

Tolstoy, Lyof, 197, 222
Tom Jones, 170-71
Torneo Notturno, 225-26
Tosca, 75, 79-83, 174, 194, 223
Tote Stadt, Die, 109
Transatlantic, 302
Traviata, La, 11-12, 70, 86, 97, 329
Tre Commedie Goldoniane, 151
Treemonisha, 175-76, 178
Trial of Lucullus, The, 260
Tristan und Isolde, 13, 32, 52, 80, 95
Trittico, Il, 87-88
Troilus and Cressida, 277-78, 279
Troyens, Les, 10, 94
Tsarskaya Nyevyesta, 31
Turandot (Busoni), 14, 46-48, 88-90,
    130, 131, 195
Turandot (Puccini), 47, 80, 88-90,
    94, 105, 173
Turn of the Screw, The, 289, 294
Twain, Mark, 308

Ulisse, 228

Vanessa, 309-10
Vanna Lupa, 224
Varney, Louis, 64
Vaughan Williams, Ralph, 171-72,
    276
Věc Makropulos, 173, 187-89, 204
Vedova Scaltra, La, 91
Vera Violanta, 109
Verdi, Giuseppe, 11-12, 15, 62, 70,
    71, 80, 95, 130, 159, 166, 171,
    183, 208, 294, 298
Vérnász, 207
Verurteilung des Lukullus, Der, 259-
    60
Vicissitudes de la Vie, Les, 189
Victory, 285-86
Vida Breve, La, 68
Village Romeo and Juliet, A, 30, 275
Vittadini, Franco, 223
Vives, Amadeo, 173
Voina i Mir, 195, 197-98
Voix Humaine, La, 219
Voják a Tanečnice, 189
Volo di Notte, 227-28
Von Heute auf Morgen, 139-40, 223

Vortice, Il, 229

Votre Faust, 221-22, 271, 339

Výlety Páně Broučkovy, 183-85, 208, 303

Wagner, Richard, 13, 14, 15, 26-27, 30, 32, 33, 34, 36, 38, 39, 50-53, 71, 80-81, 106-8, 112-14, 131, 154, 155, 158-59, 162, 182, 186, 194, 208, 219, 221, 235, 238, 249, 278, 289, 291, 294, 330, 333, 346

Wagner, Siegfried, 106-7

Wagner-Régeny, Rudolf, 254

Walton, William, 140-41, 277-78

Wandering Scholar, The, 275

War and Peace. See *Voina i Mir*

Ward, Robert, 309

Wedekind, Frank, 248, 258, 309

We Have Come to the River, 271-72

Weill, Kurt, 126, 138, 144-48, 150, 162, 174, 196, 215, 245, 247, 259, 317, 319, 320-21

Weinberger, Jaromir, 189

Weisgall, Hugo, 309

Weishappel, Rudolf, 258

Werle, Lars Johann, 204-5

Whiting, John, 192, 284

Wilbur, Richard, 320

Williamson, Malcolm, 283, 286

Willie the Weeper, 181

Wolf-Ferrari, Ermanno, 90-93, 126, 151, 153, 169-70, 233

Wolzogen, Ernst von, 112-13

Wozzeck, 80, 131, 137, 138, 150, 158-62, 173, 199, 201, 247, 248, 284, 312, 325, 326, 328, 334, 336

Wunder der Heliane, Das, 110

Wundertheater, Das, 263, 266

Wuthering Heights, 311

Zandonai, Riccardo, 78, 94, 99-102, 104, 129, 150, 173, 222

Zápisník z Mizelého, 141-42

Zauberflöte, Die, 10, 35, 121, 299, 318, 321

Zazà, 76

Zeffirelli, Franco, 310

Zerissene, Der, 262

Zimmermann, Bernd Alois, 160-61, 333-37

Z Mrtvého Domů, 187-89, 204

Zolotoy Pyetushok, 33-35, 48, 193, 196, 202

Zweig, Stephan, 237-39